PLUMBER'S LICENSING STUDY GUIDE

PLUMBER'S LICENSING STUDY GUIDE

THIRD EDITION

Michael Frankel, CPD
R. Dodge Woodson

New York Chicago San Francisco Lisbon London Madrid
Mexico City Milan New Delhi San Juan Seoul
Singapore Sydney Toronto

The McGraw·Hill Companies

McGraw-Hill books are available at special quantity discounts to use as premiums and sales promotions, or for use in corporate training programs. To contact a representative, please e-mail us at bulksales@mcgraw-hill.com.

Plumber's Licensing Study Guide, Third Edition

1 2 3 4 5 6 7 8 9 0 QDB/QDB 1 8 7 6 5 4 3 2

ISBN 978-0-07-179807-5
MHID 0-07-179807-2

This book is printed on acid-free paper.

Sponsoring Editor
Larry S. Hager

Editing Supervisor
Stephen M. Smith

Production Supervisor
Richard C. Ruzycka

Acquisitions Coordinator
Bridget L. Thoreson

Project Manager
Patricia Wallenburg, TypeWriting

Copy Editor
James Madru

Proofreader
Claire Splan

Composition
TypeWriting

Art Director, Cover
Jeff Weeks

About the Authors

Michael Frankel, CPD, is a recognized and often-quoted authority on the design and code approval of piping facilities built throughout the country and world and has acquired an intimate knowledge of many of the plumbing codes used in the design of buildings. He has been a consultant on all phases of plumbing design and engineering. Mr. Frankel, a native of Brooklyn, New York, received an engineering degree from the City University of New York and has more than 45 years of experience as a designer and department head. He is an active member of the American Society of Sanitary Engineering and the American Society of Plumbing Engineers (ASPE), and past president of the New Jersey chapter of ASPE. He is also a member of the National Fire Protection Association (NFPA) and a principal voting member of the Technical Committee on Piping Systems (HEA-PIP) for NFPA 99, *Health Care Facilities Code.*

Mr. Frankel is a prolific writer and a frequent contributor to *Plumbing Engineer, Plumbing Systems and Design,* and *PM Engineering* magazines. He is the author of *Facility Piping Systems Handbook,* now in its Third Edition (2010), and *Facility Site Piping Systems Handbook* (2012). Other accomplishments include writing the revised plumbing section for *Architectural Graphic Standards* for the American Institute of Architects and authoring multiple chapters for major handbooks such as *Piping Handbook, Practical Plumbing Engineering,* and *Handbook of Utilities and Services for Buildings.* Mr. Frankel has been a popular speaker at past ASPE conventions and symposiums and various local chapters of ASPE.

R. Dodge Woodson is a master plumber and master gasfitter with over 30 years in the plumbing business. He has owned and operated plumbing and construction companies in Virginia and Maine, and has served as adjunct faculty at the Central Maine Technical College, where he taught code and apprenticeship classes.

Contents

Introduction

All code is written to ensure that plumbing work is installed with the basic provision of safety and includes correctly designed, adequately installed, and properly maintained systems. I want to provide to students an understanding of the intent of the code and a background that will help to explain what many of the code provisions are expected to accomplish. These basic principles are not part of the code, and the definitions are only for the guidance of students. Each chapter will include a basic description of what the chapter is about and general comments that will make the material more understandable.

The purpose of this study guide is not to teach plumbing design but rather to prepare you to take and pass the journeyman or master plumber's licensing test and to answer questions similar to those given on the test. This guide will not replace the plumbing codes themselves but gives you the necessary information on how to read and interpret questions on code facts. The purpose of the test is to determine whether an individual has the necessary understanding and knowledge of various statutes in the plumbing codes and the regulations for him or her to work in any state. The knowledge gained by passing the test will allow an individual to train plumbers under his or her supervision.

Actually, reading the entire code in order to pass the licensing test is a very boring task, and besides, it is impossible to remember all aspects of it. However, if you wish to become a licensed plumber, reading and understanding the code are mandatory. It is necessary to read the code again, again, and yet again. The more you read it, the more it will become familiar. I will attempt to make this task more interesting by presenting the entire code in a logical and informative manner. This study guide will present the facts that will enable students to learn and apply the knowledge gained to answer the test questions correctly and to pass the test successfully.

Passing a plumbing licensing examination provides an opportunity for a person to enter a profession that is very satisfying. This also means not depending on a second party to obtain work. A second consideration for finding work is that once the required license is obtained, organizations that require the services of plumbers who are familiar with the plumbing code can be certain that the holder of a license is competent.

Various respected model codes, each considered reasonably complete with the most desirable aspects, are available. Such codes include

1. The *Uniform Plumbing Code* (*UPC*), sponsored by the International Association of Plumbing and Mechanical Officials (IAPMO)
2. The *International Plumbing Code* (*IPC*), sponsored by the International Code Committee (ICC)
3. The *National Standard Plumbing Code* (*NSPC*), sponsored by the National Association of Plumbing, Heating, and Cooling Contractors (NAPHCC)

Each of these codes is sponsored and endorsed by respected and knowledgeable organizations but with different interests and memberships. This leads to differences in code requirements, each organization thinking that its code is the best and most complete. This study guide will concentrate on provisions of both the *UPC* and *IPC*, which will allow readers to review questions taken from both codes. This will provide the greatest opportunity to answer questions representative of codes that cover more than 95 percent of the country.

Some parts of the code and code language are confusing because they are not written very clearly and will require interpretation. Some of the questions on the test use "weasel" words to confuse the test taker or other phrases that the test writer uses to divert your attention away from the right answer. There are also unforeseen situations that are not listed and are not covered specifically in this code. When an authority having jurisdiction (AHJ) is called on to interpret a portion of the code, the plumber should rely on basic principals to understand new or different systems and information. Deviation from the code is permitted when approved by the AHJ. Deviation indicates an alternative to what is often called a regular plumbing system. Each system must be designed by a professional engineer, understood by the installing plumber, and accepted by the AHJ.

I must point out that many of the various states, cities, and local governments throughout this country will adopt a particular code as primary and then amend it with material applicable to the local area. In general, when all the codes are compared, the differences in major sections will be mostly, but not all, small items. Because of these amendments, it is imperative to contact local officials in the area where one obtains a plumbing license and get a copy of the local code for reference and study. I emphasize that each student should obtain a copy of the applicable local code and read and reread the code again and again.

This book is intended to prepare applicants for a test that is given throughout the country. Every jurisdiction where a test is given has the prerogative to add any question that it feels is applicable to the locale. Journeyman and master plumbers have to deal with problems that are considered administration, so one should expect to find some questions about administration on the test.

Licensing exams for plumbers vary in complexity and content. In fact, they are often changed from time to time to minimize the possibility of cheating. Since I cannot predict with absolute certainty what will be on your licensing exam, this book will test you on a large part of the material that you might encounter. In all probability, the tests that I give you in each chapter will be more difficult than what you will encounter on the licensing exam. Of course, there is no way of knowing exactly what you will face on the day of the test, so it is better to be overprepared than to be underprepared.

Most of the illustrations, tables, and charts that appear in this book and are not attributed are reproduced with permission of and copyrighted by the International Code Council (IPC); all rights reserved. Other illustrations, tables, and charts taken from the *Uniform Plumbing Code* (*UPC*) are attributed to the IAPMO/American National Standards Institute (ANSI). Specific rainfall data (Figures 9.2, 9.3, and 9.4) have been obtained from the National Weather Service, National Oceanic and Atmospheric Administration (NOAA). Some of the figures are courtesy of the American Society of Plumbing Engineers (ASPE). Other figures have been taken from published articles written by me.

I would like to acknowledge Mario Berrios, P.E., plan examiner, Miami-Dade Building Department, for his help in the preparation of Chapter 5.

MASTER AND JOURNEYMAN PLUMBER EXAMINATION AND LICENSING

A master plumber is a knowledgeable and experienced person with advanced installation skills that allow him or her to work on any size or type of plumbing system. A master plumber's job requirements are more extensive than those of a journeyman, allowing the supervision of other plumbers in addition to evaluating work orders, coordinating plumbing work with the other trades, and delegating work orders to other employees. The master plumber has the ability to teach and employ plumbers at the journeyman level.

QUALIFICATIONS FOR TAKING THE TEST

There is no standard test for advancement in the plumbing profession. Every state, county, and city has a different test. Anyone thinking of taking a test is assumed to already know what the plumbing profession does. It is my purpose to include enough information to enable applicants to pass the test. Although license requirements and formats vary across the country, most areas require some sort of license. The amount of work experience that any individual must have to obtain a license differs across the states, but most states require at least four and sometimes five years of experience as a trainee or apprentice.

The first step to becoming a licensed plumber is to get recognized experience in the plumbing field. This can be accomplished by being a trainee or apprentice, which will allow you to gain the required work experience at least for the minimum length of time required by the authorities. Apprentices and trainees do not need any certification and are permitted to be hired off the street. There is usually an age requirement of 17 or 18 years of age. Some jurisdictions require a high school diploma or GED certificate to start work. There also may be a requirement in some states for an applicant to attend a trade school or to have taken other academic courses for a certain period of time. Some states do not require that an individual become certified or licensed to work as a plumber at the journeyman level. However, it is important to document your work experience.

There are two tests that are the subject of this book: the journeyman plumber's license test and the master plumber's license test. The journeyman exam is a state- or area-specific examination that will be used to determine whether the individual has the necessary skills and knowledge to work as a licensed plumber and be allowed to repair, install, and maintain any plumbing system. The standard amount of experience varies, but it is usually between four and five years as a trainee or apprentice. This requires the individual to work under the supervision of a master plumber for that time. Individuals who are not required to have a journeyman's license but have met all the other requirements set by the state to work at the journeyman's level still can do the same work. Journeymen are not usually permitted to manage plumbing projects that employ a large number of other plumbers.

The master exam is a state- or area-specific examination that is similar to the journeyman exam. The normal amount of work experience necessary is at least four to five years of experience as a journeyman plumber. Master plumbing job requirements are more extensive than those of a journeyman. It is the responsibility of a master plumber to evaluate work orders, coordinate and delegate work orders to other employees, train and manage employees, and meet with potential clients to review plumbing problems. It is often a requirement to have knowledge of federal plumbing codes as well as the financial basics of owning a business.

As a final word, it is imperative that you check the laws, rules, and regulations in the particular jurisdiction where you wish to take the test.

Michael Frankel, CPD

Definitions

Definitions will not be part of the test. This partial list is provided to clarify any term used in the code that might be confusing. The definitions have been paraphrased to make them more understandable and less likely to be misinterpreted, as well as to make them usable for more than one code.

Accessible or readily accessible *Accessible* means "being easily approachable" such as to an appliance, valve, fixture or other equipment that might require removal or movement of a door, panel, or other obstruction. Readily accessible means direct approachability to an appliance, valve, fixture, or other equipment be available without the need to remove or move any obstruction.

Air break An arrangement of piping where a drain from a fixture, appliance, or other equipment discharges indirectly into another fixture below the flood level of that fixture but above the trap seal.

Air gap The unobstructed vertical distance through the air between the lowest opening of any pipe and the flood-level rim of the receptacle or tank into which the pipe is discharging.

Alternative engineering design A plumbing system not specifically regulated by a code that performs in accordance with the intent of the code and provides an equivalent level of protection of public health, safety, and welfare.

Approved Acceptable to the code official or other AHJ.

Authority having jurisdiction (AHJ) The organization, office, entity, individual, or others having statutory authority for enforcing the requirements of a code or standard. This also includes a duly authorized representative of that authority.

Backflow The reverse flow of water, contaminated water, or other liquids into a potable water distribution system other than from its intended source, caused by an excess of pressure.

Backsiphonage The reverse flow of contaminated water or other liquids into a potable water distribution system as a result of the pressure in the potable water system falling below atmospheric pressure, causing a vacuum.

Building drain That part of the lowest portion of the drainage system inside a building that receives the discharge from all soil, waste, and other drainage piping. The building drain extends beyond the building wall onto the site for a distance regulated by local authorities. The building drain conveys sanitary sewage only or storm water and other drainage only but no sewage. A combined drain conveys both sanitary and storm water drainage.

Building sewer That part of the horizontal drainage system outside a building that extends from the end of the building drain and conveys the discharge of the building drain to the point of ultimate discharge, such as a public sewer.

Code The regulations and standards that are a compilation of provisions covering a broad subject matter that the administrative authority has lawfully adopted and is independent of other codes and standards.

Code official The designated officer or other entity or a duly authorized representative charged with enforcement and administration of a code.

Cross-connection The physical connection between a potable water supply and any other separate piping system or equipment where it might be possible for a nonpotable liquid to flow from one system to a part of the potable water system under any condition.

Developed length The length measured along the centerline of a pipe or fitting.

Existing work Any plumbing system regulated by the code in use that was legally installed prior to adoption of the particular code.

Fixture unit This is a measure of the probable usage or discharge by various types of plumbing fixtures and equipment based on some arbitrarily selected scale. The fixture unit value depends on the volume rate of drainage discharge (drainage fixture unit) or usage rate of water (water fixture unit) and also considers duration of a single operation and the average time between successive operations.

Flood level A rim or edge of a fixture or receptacle above which water overflows.

FOG Fats, oils, and grease.

Fuel gas Natural, manufactured, or liquefied petroleum or a mixture of these.

Gray water Waste discharged from lavatories, bathtubs, and showers and/or clothes washers and laundry trays that has not come into contact with any toilet waste.

Grease interceptor and removal device A device designed and installed in a sanitary drainage system for waste discharged from kitchens to provide a method to separate free-floating grease from the drainage stream by gravity. Once the grease is removed, the normal waste is allowed to be discharged into the drainage system. This device could be a large interceptor using only gravity or a smaller hydromechanical interceptor that uses baffles and air entrainment to separate out the grease. The removal device mechanically removes the accumulated fats, oils, and grease (FOG) either by an automatic or by a manual process.

Indirect waste pipe A drainage pipe that does not connect directly with a drainage system but rather discharges into another receptacle that is connected directly to the drainage system.

Plumbing The business, trade, or work having to do with the installation, removal, alteration, or repair of sanitary or storm drainage facilities, venting systems, public and private water supply systems, and gas systems.

Plumbing fixture An approved installed receptacle, device, or appliance that is supplied with water or that receives liquid or liquid-borne waste and discharges such waste into a drainage system to which it may be connected directly or indirectly. Processing equipment is not a plumbing fixture.

Sewage Liquid waste containing human, animal, or vegetable matter in suspension or solution and may include chemicals.

Trap A fitting or device designed and constructed to provide, where properly vented, a liquid seal that prevents the passage of sewer gases without restricting the flow of sewage or wastewater through it.

Vent A pipe or pipes installed to provide a flow of air to or from a drainage system to provide a circulation of air within such a system to protect trap seals and to equalize air pressure.

Waste pipe A pipe that conveys only liquid waste, free of fecal matter.

PLUMBER'S LICENSING STUDY GUIDE

Chapter 1
ADMINISTRATIVE POLICIES

This chapter addresses requirements that pertain to buildings and structures included in the body of the code as well as regulations that relate to code enforcement in an equitable manner. Every jurisdiction that administers a test has the prerogative of adding any question that it feels is appropriate for the locale. Journeyman and master plumbers have to deal with problems that are considered administration, and therefore, one should expect to find some questions about administration on the test. The rules for master plumbers are more complex and comprehensive than they are for journeyman plumbers, so questions addressing such rules may not be present on both tests.

In addition to the many specific aspects of the plumbing codes, test takers are expected to know a fair number of general code regulations. These general elements of the code are just as important as any of the specific items in a licensing exam. Not only are the general regulations required reading for the test, but many of them are used frequently in the field. It is important to learn these general regulations for the exam.

You might say that the general regulations are the foundation on which the entire code is built. Putting it in terms of construction, you could say that the general regulations are the footing for the whole structure. If you know anything about construction, you know how critical a good footing is. Just as a footing is of paramount importance in construction, the general regulations of the plumbing code are equally important in application of and compliance with the code.

What types of requirements do the general regulations cover? The list of subjects is not particularly long, but it is important. For example, the general regulations deal with health and safety. As a plumber, you might be thinking that this has to do with potable water and sewage. In reality, the general regulations go much deeper than that. Some of the regulations may not seem to be specific to plumbing, but they are related to work that is done in conjunction with the installation of plumbing.

MULTIPLE-CHOICE EXAM

1. When there are conflicts between referenced standards, referenced codes, and other legally applicable provisions, which of the following takes precedence?

   a. The referenced code
   b. The referenced standard
   c. Other applicable provisions
   d. Provisions in the code

2. When items that pertain to either a new or existing building are essential for health and safety and are not covered in the code book, who has the authority to make a decision on how a matter should be resolved?

   a. A master plumber
   b. A journeyman plumber
   c. A code official
   d. None of the above

3. When is a repair to an existing system not exempt from requiring a permit?

   a. When it involves stopping a leak in a concealed trap
   b. When it involves stopping a leak in a defective water line
   c. When it involves stopping a leak in a broken drain line
   d. When it involves removing and replacing a water closet

4. If an alternative material is submitted for approval, who is responsible for the approval?
   a. The master plumber
   b. The property owner
   c. The applicant of the request
   d. The code-enforcement office

5. Who is responsible for code-compliance inspections of plumbing systems?
   a. The master plumber
   b. The property owner
   c. The code official
   d. The fire marshal

6. When a plumbing permit is required for proposed work, when may the work begin?
   a. When a permit is applied for
   b. When a permit is issued
   c. When a permit is approved
   d. When a permit is signed by a code official

7. Which of the following types of work requires a plumbing permit?
   a. The replacement of a water heater
   b. The repair of a pipe that froze and split
   c. The replacement of an existing valve
   d. The cleaning out of a stopped drain

8. A plumbing permit is *not* valid until it is
   a. posted in a conspicuous place on the job site.
   b. posted in the owner's office on site.
   c. signed by the master plumber or an authorized representative.
   d. signed by the code official or an authorized representative.

9. A nonemergency stop-work order is *not* valid if it is given to
   a. the owner of the property.
   b. a mechanic doing the work.
   c. the owner's agent.
   d. the master plumber verbally.

10. When will a code officer issue a notice of approval for a plumbing system?
    a. When all the work is complete
    b. When all tests and inspections ensure compliance with the code
    c. When the owner requests it
    d. When requested to do so in writing

11. A code official may extend the time allowed for a permit application to be effective by a period of how many days?

a. 30 b. 60

c. 90 d. 120

12. Any plumbing permit issued may become invalid if work is not started for how long?

a. 90 day b. 180 days

c. 270 days d. 360 days

13. Which of the following are required to be supplied with a set of approved construction documents?

a. The code-enforcement office b. The job site

c. The engineer's office c. The owner's office

14. Which of the following is a suitable cause for requesting an appeal?

a. When a provision of the code does not apply

b. When a better building form is proposed

c. When the code has been interpreted incorrectly

d. When a code requirement should be waived

15. Who or what has jurisdiction over the supply of a temporary potable water supply?

a. Building code official

b. The plumbing codes

c. The local town administrative office

d. The town water purveyor

16. A permit application need *not* include which of the following information?

a. A full description of the plumbing

b. The name of the general contractor

c. A set of detailed plans

d. The name of the owner

17. When may a request for an inspection of plumbing work be made?

a. After covering the underground site work

b. After installation of the ceiling membranes

c. On request by the contractor

d. Prior to roughing in the sanitary piping

18. A code official is appointed by whom?

a. The commissioner of buildings
b. The director of plumbing inspection
c. The jurisdictional attorney
d. The chief appointing authority or AHJ

19. What is the primary purpose of a written contract?

a. To define the obligations of each party
b. To standardize work performance
c. To condense construction performance
d. To define codes and standards

20. If a part of a building is in violation of the code and a code officer condemns the building, under what conditions can the building continue to be used without violating the code?

a. No conditions
b. Not permitted
c. With a written agreement
d. When corrections are made in 14 days

21. When is an alteration or repair to an existing system *not* in violation of the plumbing code?

a. When it is done without a permit
b. When it is not done by a master plumber
c. When the plumbing system becomes unsafe
d. When a previously approved pipe is replaced

22. The standards referenced in the code are to be considered

a. construction alternatives.
b. mandatory requirements.
c. a list of material substitutions.
c. material modifications.

23. Under what circumstances will a temporary connection be allowed by a code official?

a. When it removes a violation
b. When it is to be used for a period of 90 days or less
c. When it is requested by the property owner
d. When it is used to test the plumbing system

24. The plumbing code governs which of the following?

a. Sanitary drainage piping
b. Fuel-gas piping
c. Medical oxygen systems
d. Detached multiple dwellings

25. Which of the following is *not* the primary intent of the plumbing code?

a. To allow revenue collection for the local jurisdiction

b. To allow the approved installation of plumbing piping

c. To regulate plumbers

d. To ensure public safety, health, and welfare

26. What plumbing repairs must comply with or conform to new code provisions?

a. Existing work that includes no occupancy change

b. The clearing of stoppages

c. Work that requires replacement of piping

d. The repair of leaks

27. If a building is scheduled to be demolished, who must be notified of the intended work?

a. A master plumber

b. The property owner

c. The code official

d. The utility companies with connections

28. Which of the following is *not* the intent of the plumbing code?

a. To allow the local jurisdiction to levy fees

b. To ensure public safety and health

c. To regulate plumbers

d. To regulate the design and construction of a project

29. The information and standards appearing in the appendices of the code do *not* apply unless

a. they are relevant to fire or life safety.

b. they are referenced in the code.

c. they are specifically adopted.

d. they are applicable to specific conditions.

30. Who is required to enforce the plumbing code?

a. A master plumber

b. A person appointed by the code official

c. The owner or his or her representative

d. A person appointed by the owner

31. When a manufacturer's installation requirements are different from code requirements, which should be used?

a. The code official's requirements

b. The manufacturer's instructions

c. The referenced standards

d. The most restrictive requirements

32. What types of facilities must be provided for workers on construction sites?

a. Recreational facilities
b. Toilet facilities
c. Rest facilities
d. A means of heating

33. When altering a structure for the installation of plumbing, the job must be left in what condition?

a. A safe and nonhazardous condition
b. A condition that is equal to the way it was found
c. A neat and clean condition
d. A condition that is satisfactory to the property owner

34. When a structural hazard exists on a premise, who is responsible to abate such a nuisance?

a. A code official
b. A master plumber
c. The property owner or an authorized agent
d. The police department

—NOTES—

Chapter 2
MISCELLANEOUS POLICIES

This chapter addresses questions that do not fall into any one particular category and therefore are not specific to any of the other chapters but are necessary for understanding the code. In addition, this chapter addresses questions regarding materials, installation, supports, and other topics that do not fall into specific categories and may be appropriate to other areas of the code.

In addition to the many specific aspects of the plumbing code, there are a fair number of general code regulations that you must know. These elements of the code are just as important as any other in a licensing exam. Not only are the general regulations required reading for the test, but many of them are used frequently in the field. For example, the general regulations dictate how the testing of a plumbing system must be done

When you read the general regulations, you will notice provisions that deal with cutting and notching structural members when you need to run a horizontal drain through floor joists and you cannot keep the pipe in the center of the joists. Since the drain must be graded properly, you may find that your holes get closer and closer to the bottom of the joist as you go along. How close to the bottom of a joist can you drill a hole without causing structural or code problems? If you are in charge of the plumbing job, you need to know the answer to the question. Drilling holes too close to the bottom of a joist can result in an expensive and time-consuming structural repair for which you may be held responsible.

For example, there is a code requirement stating that a trench shall not be installed closer than a 45-degree-angle line to the load-bearing plane of the footing. The reason for this rule is fairly obvious: A more severe angle line for the trench could weaken the structural integrity of the footing. It is possible that you may never need to use this or certain other requirements found in the general regulations, but you never know what might come out of left field when you least expect it.

MULTIPLE-CHOICE EXAM

1. A third-party certification is required for which of the following piping systems?

a. A potable water system
b. A sanitary drainage system
c. A special waste system
d. All plumbing products

2. Waste water condensate piping from cooling coils shall be pitched at which of the following slopes?

a. $\frac{1}{16}$ inch per foot (0.5 cm/m)
b. $\frac{1}{8}$ inch per foot (1 cm/m)
c. $\frac{1}{4}$ inch per foot (2 cm/m)
d. $\frac{1}{2}$ inch per foot (4 cm/m)

3. Where shall plumbing pipes subject to freezing be installed?

a. Within walls of a building
b. In an attic
c. In a crawl space
d. Within an area protected by insulation

4. Any pipe that passes under a footing or through a foundation must be protected by a relieving arch or which of the following?

a. A pipe sleeve
b. Tie rods
c. A backwater valve
d. An expansion joint

5. Exposed piping in plenums shall comply with which of the following codes?

a. *International Fire Code*
b. *International Mechanical Code*
c. *International Plumbing Code*
d. *International Fuel Gas Code*

6. Which type of pipe does *not* require the installation of nail plates to protect it when passing through holes or notches in studs?

a. Polyvinyl chloride (PVC)
b. Acrylonitrile butadiene styrene (ABS)
c. Copper
d. Cast iron

7. What is the minimum thickness required for a nail-plate installation?

a. 0.035 inch (0.89 mm)
b. 0.057 inch (14.48 mm)
c. 0.062 inch (1.58 mm)
d. 0.075 inch (19.05 mm)

8. At what interval between supports shall buried pipe be suspended?

a. At 4-foot (1.20-m) intervals
b. On blocks to grade
c. At 10-foot (3.0-m) intervals
d. On solid supports between bell holes for the entire length except for bell hole

9. When shall trenches containing pipe *not* be completely backfilled?

a. When loose backfill has been placed around the pipe
b. When the backfill material is dry
c. When rocks have been removed from the backfill
d. When the pipe has been tested

10. Notches on the ends of joists used for installation of pipe must *not* exceed which of the following requirements?

a. Those of the property owner
b. The plumbing code
c. Those of the master plumber
d. The *International Building Code*

11. The annular space between the exterior of a pipe and the inside of a pipe sleeve is not permitted to be sealed with which of the following materials?

a. Caulking material
b. Backfill material
c. A casketing system
d. Foam sealant

12. Separate plumbing facilities for each gender shall be provided for which of the following situations?

a. Dwelling and sleeping units
b. Tenant spaces with fewer than 15 people
c. Listed occupancies in required fixtures
d. Mercantile facilities with no more than 50 people

13. A plumbing fixture that discharges only clear waste, such as a kitchen sink, can be drained primarily into which of the following?

a. A sanitary drainage system
b. A French drain
c. A gravel-lined pit
d. A system intended to flush toilets

14. Pipe installed in trenches dug parallel to the footing of a structure must be extended below which of the following?

a. A 22½-degree angle of the bearing plane
b. A 45-degree angle of the bearing plane
c. A 60-degree angle of the bearing plane
d. A 90-degree angle of the bearing plane

15. Plumbing piping is prohibited from being installed in elevator shafts and elevator equipment rooms *except* for which of the following?

a. Floor drains
b. Sumps and sump pumps
c. Piping in hydraulic elevator pits
d. Where an oil separator is installed

16. When backfilling a trench, the fill material must be installed in layers. How thick should each layer be?

a. 6 inches (152.40 mm)
b. 4 inches (101.6 mm)
c. 8 inches (203.3 mm)
d. 10 inches (254 mm)

17. It is *not* necessary for installed piping to be protected against which of the following:

a. Expansion
b. Stresses
c. Freezing
d. Water damage

18. Which of the following materials is unsuitable for piping backfill when a trench has been overexcavated?

a. Fine gravel
b. Crushed stone
c. Concrete
d. Rock removed from the excavation

19. When a trench is backfilled, at what intervals must the fill material over the pipe be compacted?

a. Every 4 inches (101.6 mm)
b. Every 6 inches (152.4 mm)
c. Every 8 inches (203.2 mm)
d. Every 12 inches (304.8 mm)

20. Which of the following types of pipe do *not* require the installation of shield plates to protect them?

a. Polyvinyl chloride (PVC)
b. Acrylonitrile butadiene styrene (ABS)
c. Copper
d. Cast iron

21. Pipes in trenches may *not* be covered with backfill until

a. the pipe has been approved by a code official.
b. the backfill material is dry.
c. the pipe has been tested for stress.
d. approved by the health department.

22. When drilling holes in joists for the passage of pipes, the holes must *not* be drilled closer than

a. the general contractor permits.
b. the building code allows.
c. the fire code allows.
d. the plumbing code allows.

23. Notches on the ends of joists used for piping installation must not exceed the requirements of

a. the property owner.
b. the general contractor.
c. the building code.
d. the *International Building Code.*

24. Gauges used for plumbing system test purposes at 10 pounds per square inch (psi) or less, shall have which of the following minimum increments?

a. 1 psi (0.07 kg/cm^2)
b. 0.5 psi (0.035 kg/cm^2)
c. 2 psi (0.14 kg/cm^2)
d. 0.10 psi (0.007 kg/cm^2)

25. By what method shall plumbing fixtures discharge into a sanitary system?

a. Direct
b. Indirect
c. Separate
d. An air gap

26. Where the frost line is established at 30 inches (750 mm) below grade, how deep must a water pipe be installed?

a. 18 inches (457 mm) below the frost line

b. 12 inches (305 mm) below the frost line

c. 6 inches (227 mm) below the frost line

d. 18 inches (457 mm) below grade

27. The one criterion *not* required for an alternative engineered design is which of the following?

a. Being designed by a registered design professional engineer

b. Submission of signed and sealed construction documents to the code official

c. Sufficient technical data to substantiate performance of the proposed design

d. Being designed and approved by a master plumber

28. When testing a drain and vent system with water, what is the minimum head under which the system is tested?

a. 5 feet (1.5 m)

b. 10 feet (3.0 m)

c. 15 feet (4.5 m)

d. 12 feet (3.6 m)

29. The materials and cost of inspection and testing for plumbing systems by a permit holder shall include which of the following?

a. The code official's cost

b. The permit holder's fee

c. The cost to the owner of the property

d. The master plumber's fee

30. Air is an accepted means to test a piping system made up of which of the following piping materials?

a. Plastic piping

b. Copper piping

c. Cast-iron piping

d. Galvanized piping

31. A drainage and vent system air test shall be held for how long?

a. 5 minutes

b. 10 minutes

c. 15 minutes

d. 20 minutes

32. How long is water required to remain in a drainage system prior to an inspection of the joints?

a. 10 minutes

b. 15 minutes

c. 20 minutes

d. 30 minutes

33. If water is used to test water-distribution piping, where must one obtain the test water?

a. Any water well
b. A clear-flowing stream
c. A potable source
d. A clear lake

34. If air is used to test a drain, waste, and vent system, what is the required test pressure?

a. 5 psi (0.35 kg/cm^2)
b. 10 psi (0.70 kg/cm^2)
c. 15 psi (1.05 kg/cm^2)
d. 18 psi (1.26 kg/cm^2)

—NOTES—

Chapter 3
WATER SUPPLY AND BACKFLOW PREVENTION

There is a great deal to learn about the subject of potable water systems and backflow prevention. The rules and regulations pertaining to potable water supplies and backflow affect all plumbers.

The IPC code, Section 608.1, states that the potable water supplies shall be "designed, installed and maintained in such a manner that will prevent contamination from nonpotable liquids, solids or gases being introduced into the potable water supply through cross-connection or any other piping connections to the system." This is accomplished through backflow-prevention applications.

Potable water, of course, is another name for drinking water. Good, clean water is essential to life, and it is the job of plumbers to install systems that supply reliable potable water. If plumbers make mistakes in their installation methods or material choices, serious health problems can result for users of the system. If you make a mistake that creates a health problem for the person for whom you installed the system, you could be open to legal action or your employer could be sued. What's worse, someone could suffer from illness, serious burns, or even death as a result of your mistake. Potable water is not a subject to be taken lightly.

Backflow can take either of two forms, and it is critical to prevent both. The first is backflow that originates from the drainage system and flows into a potable water distribution piping. Prevention of this type of backflow is covered in Chapter 7. The other form is backflow into a potable water system that originates from the water piping distribution system and is not connected to the drainage system. This category of backflow is the result of cross-connections and vacuum that is created by accident. This happens when a potable water outlet (such as a flexible hose) is submerged into a contaminated source and an accident allows a potential reverse flow. Both types of backflow preventers that have accepted devices to prevent backflow and provide safety to the potable water system. Table 3.1 matches the various backflow preventers with the degree of hazard in a particular system. The following figures and explanations will provide applicants with the important features of the various types of backflow preventers.

1. The *degree of hazard* (high or low) is determined by the code official. The required protection (backpressure or backsiphonage) is determined by job requirements. The correct backflow preventer is often determined by the licensed plumber, who selects the device based on the hazard and application. A complete listing of selection criteria is provided in Table 3.1.
2. *Backpressure* occurs only in a continuously pressurized piping system where there is water pressure on both sides of the point in question and flow is intended to be in only one direction. Any increase in the downstream pressure at the point in question will cause an increased upstream pressure in the direction of flow. An example of backpressure is seen when the pressure in a boiler exceeds the makeup water piping pressure, thus forcing water from the boiler into the potable water supply. For protection from this type of backflow, one uses a large and costly reduced-pressure principal backflow preventer. This device is considered the most reliable form of protection outside of an air gap. This type of preventer is shown in Figure 3.1.
3. *Backsiphonage* is the flow of water where there is no continuous pressure. This occurs when the pressure at some point in a water distribution system drops below the normal pressure in a part of the system. An example of backsiphonage would be seen when there is a break in the main or public water supply that causes a loss of pressure. The loss of pressure causes siphonic action to occur. Water upstream of the break would be siphoned back to the point of the break, creating a vacuum in the piping. For this type of backflow, protection from a double-check-valve assembly or vacuum breaker is considered adequate. The double-check-valve is shown in Figure 3.2.

4. A nonpressurized vacuum-breaker backflow device is used to prevent contaminated water from entering a potable water system through such fittings as a hose bibb or other fixture that could have an attached flexible hose. The problem occurs when a flexible hose is submerged in a puddle of contaminated water. If pressure is lost upstream, a siphonic (vacuum) action occurs. This device allows air to enter an air inlet port, breaking the vacuum and preventing the backflow of water. A typical nonpressurized vacuum-breaker backflow preventer is illustrated in Figure 3.3.

There will be numerous questions on the licensing exam that pertain to potable water. There will be more than enough questions of this type to cause you great difficulty in passing the exam if you are not well prepared with a depth of knowledge on water distribution and backflow prevention.

MULTIPLE-CHOICE EXAM

1. What is water called that is safe for drinking or culinary purposes?

   a. Potable water b. Nonpotable water
   c. Lake water d. Unapproved water

2. Nonpotable water is permitted to be used for which of the following purposes?

   a. Personal use b. Culinary use
   c. Medical use d. Car washing

3. What is the name of a device used to absorb pressure surges that occur when a flow of water is stopped rapidly?

   a. A pressure-reducing valve b. A vacuum breaker
   c. A water-hammer arrester d. A pressure-relief valve

4. Where could one locate a plumbing system valve that could be fully opened?

   a. Near a curb b. On the discharge of a meter
   c. Near a building d. At the entrance to a residential building

5. Potable water need not be supplied for which of the following uses?

   a. Drinking b. Car washing
   c. Culinary uses d. Bathing

6. What is the maximum water pressure permitted inside a building?

   a. 80 psi (5.6 kg/cm^2) b. 70 psi (4.9 kg/cm^2)
   c. 60 psi (4.20 kg/cm^2) d. 50 psi (3.5 kg/cm^2)

7. A fixture water-supply pipe is permitted to terminate at what maximum distance from the fixture?

a. 18 inches (457 mm) b. 24 inches (635 mm)
c. 30 inches (762 mm) d. 48 inches (1,219 mm)

8. Hot water is *not* required in a residential building when it is used for which of the following purposes?

a. Plumbing fixtures b. Kitchen sinks
c. Building maintenance d. Equipment washing

9. Hot water is *not* required in a nonresidential building when it is used for which of the following purposes?

a. Laundry b. Culinary uses
c. Bathing d. Preheating water

10. Hot water *is* required in a residential building where which of the following is provided?

a. Bathing b. Maintenance of livestock
c. Building maintenance d. Car washing

11. Cold water is *not* required in buildings

a. where plumbing fixtures are installed.
b. used for human occupancy.
c. where livestock is maintained.
d. used for culinary purposes.

12. Potable water is *not* required for which of the following buildings?

a. Those used for food processing
b. Those used for culinary purposes
c. Those used for medical and pharmaceutical processing
d. Those used for equipment washing

13. What is the minimum diameter of a water-service pipe?

a. ½ inch (DN 15) b. ¾ inch (DN 20)
c. 1 inch (DN 25) d. 1¼ inch (DN 32)

14. A water-service pipe and a building sewer must be separated by which of the following?

a. A masonry barrier b. Undisturbed earth
c. Compacted earth d. Not less than 5 feet of compacted earth

15. The bottom of a water-service pipe within 5 feet of a sewer never should be closer than how many inches from the top of that sewer?

a. 6 inches (152 mm)
b. 12 inches (305 mm)
c. 18 inches (457 mm)
d. 24 inches (610 mm)

16. Water-service pipes may not be installed in, under, or above which of the following?

a. An approved sewer pipe
b. A septic tank
c. A septic discharge field
d. A cesspool

17. What is the main water pipe delivering potable water to a building called?

a. A water-service pipe
b. A drainage pipe
c. An acrylonitrile butadiene styrene (ABS) pipe
d. A ductile-iron water pipe

18. Under normal conditions, a water hammer arrester is required wherever certain valves are used. Which of the following valves is of a type to require such a device?

a. A gate valve
b. A stop-and-waste valve
c. A globe valve
d. A quarter-turn valve

19. When the pressure in a water main that supplies potable water to a building fluctuates, what shall be the criterion for pressure for design purposes?

a. Minimum
b. Maximum
c. Average
d. Consolidated

20. Where manifolds for potable water are installed, they must be

a. 42 inches (1,077 mm) above the finished floor.
b. accessible.
c. 60 inches (1,524 mm) above the finished floor.
d. easily identified.

21. In a drilled well, at what minimum dimension below the ground surface is cement grout required when it is used to fill around the exterior casing of the well?

a. 15 feet (4.5 m)
b. 12 feet (4 m)
c. 10 feet (3 m)
d. 5 feet (1.5 m)

22. All the following joints are prohibited on potable water piping except

a. cement.
b. concrete.
c. saddle-type fittings.
d. hub and spigot fittings.

23. What is the minimum size of a water-supply pipe to a flush-tank toilet?

a. ⅜ inch (DN 10)
b. ½ inch (DN 15)
c. ¾ inch (DN 20)
d. 1 inch (DN 25)

24. What is the minimum size of a water-supply pipe for a wall hydrant?

a. ⅜ inch (DN 10)
b. ½ inch (DN 15)
c. ¾ inch (DN 20)
d. 1 inch (DN 25)

25. What is the minimum size of a specific fixture water-supply pipe for a flush-valve urinal?

a. ⅜ inch (DN 10)
b. ½ inch (DN 15)
c. ¾ inch (DN 20)
d. 1 inch (DN 25)

26. What is the minimum size of a specific fixture water-supply pipe for a single showerhead?

a. ⅜ inch (DN 10)
b. ½ inch (DN 15)
c. ¾ inch (DN 20)
d. 1 inch (DN 25)

27. What is the minimum size of a specific fixture water-supply pipe for a lavatory?

a. ⅜ inch (DN 10)
b. ½ inch (DN 15)
c. ¾ inch (DN 20)
d. 1 inch (DN 25)

28. What is the minimum size of a specific fixture water-supply pipe for a bathtub?

a. ⅜ inch (DN 10)
b. ½ inch (DN 15)
c. ¾ inch (DN 20)
d. 1 inch (DN 25)

29. All reclaimed and gray nonpotable water shall be identified by what color?

a. Green
b. Red
c. Purple
d. Brown

30. Atmospheric vacuum breakers shall not be installed at what locations:

a. Hose connections
b. Flush valve discharge
c. Hose bibb outlets
d. Chemical dispensers

31. Connections to lawn sprinkler systems are not permitted to be protected by

a. double-check-valves.

b. pressure vacuum breakers.

c. reduced-pressure vacuum breakers.

d. atmospheric vacuum breakers.

32. For health-care facility piping, a vacuum breaker installed for equipment shall be not less than how many inches above the flood level of the fixture?

a. 12 inches (305 mm)
b. 8 inches (214 mm)
c. 6 inches (153 mm)
d. 4 inches (107 mm)

33. Where an air gap is used as protection for a potable water supply, the minimum distance from the lowest end of the outlet to the fixture or receptacle receiving discharge shall be from

a. the weir of the trap.

b. the flood level of the rim.

c. the air break inside the receiving fixture.

d. 2 inches (50 mm) above the fixture outlet.

34. Connections to a fire protection system are *not* permitted to be made with which of the following?

a. A double-check-valve assembly

b. A double-check-valve

c. A reduced-pressure backflow assembly

d. No backflow provision is necessary.

35. The overflow from a water (house) tank supplying a building should be installed to discharge how far above the roof?

a. 6 inches (50 mm)
b. 12 inches (100 mm)
c. 18 inches (150 mm)
d. 24 inches (200 mm)

36. What fixtures are *not* required to have the hot water outlet oriented to the left side of the fixture?

a. Lavatory outlets

b. American Society of Sanitary Engineers (ASSE)-rated mixing valves

c. Kitchen sink faucets

d. Bathtub outlets

37. Which of the following beverage dispensers does *not* require backflow protection?

a. A coffee machine
b. A carbonated-beverage dispenser
c. An ice machine
d. A noncarbonated-beverage dispenser

38. Potable water piping shall be disinfected with chlorine at what level?

a. 100 parts per million (ppm)
b. 75 ppm
c. 50 ppm
d. 25 ppm

39. Potable water piping with 200 ppm of chlorine that is used to achieve adequate disinfection of piping, shall be left to stand for what period of time?

a. 6 hours
b. 2 hours
c. 4 hours
d. 3 hours

40. When used, how high shall a barometric loop extend above the water connection to a fixture?

a. 40 feet (13 m)
b. 35 feet (11 m)
c. 30 feet (9 m)
d. 25 feet (7.5 m)

41. What is the hanger spacing required for 2-inch chlorinated polyvinyl chloride (CPVC) cold-water piping?

a. 3 feet (1 m)
b. 4 feet (1.2 m)
c. 6 feet(2 m)
d. 10 feet (3 m)

42. What is the maximum horizontal hanger spacing for 1½-inch and larger copper tubing?

a. 3 feet (1 m)
b. 4 feet (1.2 m)
c. 10 feet (3 m)
d. 12 feet (3.6 m)

43. If a water pump is installed in a basement, how high shall the pump be mounted above the floor?

a. 18 inches (441 mm)
b. 16 inches (406 mm)
c. 12 inches (305 mm)
d. 24 inches (610 mm)

44. What is the maximum water pressure permitted for a private lavatory?

a. 25 psi (167 kPa)
b. 50 psi (335 kPa)
c. 60 psi (402 kPa)
d. 70 psi (470 kPa)

45. Threaded joints for Schedule 80 chlorinated polyvinyl chloride (CPVC) pipe are permitted if the pressure rating of the pipe is reduced by what percentage?

a. 25 percent
b. 50 percent
c. 10 percent
d. 60 percent

46. The water supply for hospital fixtures is *not* permitted to be protected by which of the following?

a. A reduced-pressure principal backflow preventer

b. An air gap

c. An atmospheric vacuum breaker

d. A double-check-valve

47. How should mechanical joint pipe be installed?

a. Per manufacturer's instructions

b. With compression joints

c. With threaded joints

d. With grooved joints

48. Mechanically formed tee fittings shall be formed with the height of the extracted outlet at what thickness of the branch-tube wall?

a. Three times

b. Two times

c. Four times

d. The same thickness

49. What is the maximum water pressure permitted within a building?

a. 20 psi (140 kPa)

b. 40 psi (280 kPa)

c. 60 psi (420 kPa)

d. 80 psi (560 kPa)

50. When using the *UPC*, what is the maximum cold water velocity permitted in cold weather?

a. 4 feet per second (1.2 meters per second)

b. 6 feet per second (2.0 meters per second)

c. 8 feet per second (2.4 meters per second)

d. 10 feet per second (3.0 meters per second)

TABLE 3.1 Application of Backflow Preventers

DEVICE	DEGREE OF HAZARD	APPLICATION	APPLICABLE STANDARDS
Air gap	High or low hazard	Backsiphonage or backpressure	ASME A112.1.2
Air-gap fittings for use with plumbing fixtures, appliances, and appurtenances	High or low hazard	Backsiphonage or backpressure	ASME A112.1.3
Antisiphon-type fill valves for gravity water closet flush tanks	High hazard	Backsiphonage only	ASME 1002, CSA B125
Backflow preventer for carbonated beverage machines	Low hazard	Backpressure or backsiphonage Sizes ¼ to ⅜ inch	ASSE 1022, CSA B64.3.1
Backflow preventer with intermediate atmospheric vents	Low hazard	Backpressure or backsiphonage Sizes ¼ to ¾ inch	ASSE 1012, CSA B64.3
Barometric loop	High or low hazard	Backsiphonage only	See Section 608.13.4.
Double-check-valve backflow prevention assembly and double-check-valve fire protection backflow prevention assembly	Low hazard	Backpressure or backsiphonage (fire sprinkler systems) Sizes ⅜ to 16 inches	ASSE 1015, AWWA C510, CSA B64.5, CSA B64.5.1
Double-check-valve detector fire protection backflow prevention assemblies	Low hazard	Backpressure or backsiphonage (fire sprinkler systems) Sizes 2 to 16 inches	ASSE 1048
Double-check-valve-type backflow preventer	Low hazard	Backpressure or backsiphonage Sizes ¼ to 1 inch	ASSE 1024, CSA B64.6
Hose-connection backflow preventer	High or low hazard	Low-head backpressure, rated working pressure, backpressure, or backsiphonage Sizes ½ to 1 inch	ASSE 1052, CSA B64.2.1.1
Hose-connection vacuum breaker	High or low hazard	Low-head backpressure or backsiphonage Sizes ½, ¾, and 1 inch	ASSE 1011, CSA B64.2, CSA B64.2.1
Laboratory faucet backflow preventer	High or low hazard	Low-head backpressure or backsiphonage	ASSE 1035, CSA B64.7
Pipe-applied atmospheric-type vacuum breaker	High or low hazard	Backsiphonage only Sizes ¼ to 4 inches	ASSE 1001, CSA B64.1.1
Pressure vacuum breaker assembly	High or low hazard	Backsiphonage only Sizes ½ to 24 inches	ASSE 1020, CSA B64.1.2
Reduced-pressure principal backflow preventer and reduced-pressure principal fire protection backflow preventer	High or low hazard	Backpressure or backsiphonage Sizes ⅜ to 16 inches	ASSE 1013, AWWA C511, CSA B64.4, CSA B64.1
Reduced-pressure fire protection backflow prevention assemblies	High or low hazard	Backsiphonage or backpressure (fire sprinkler systems)	ASSE 1047
Spillproof vacuum breaker	High or low hazard	Backsiphonage only Sizes ¼ to 2 inches	ASSE 1056
Vacuum-breaker wall hydrants, frost-resistant, automatic draining type	High or low hazard	Low-head backpressure or backsiphonage Sizes ¾ and 1 inch	ASSE 1019, CSA B64.2.2

For SI: 1 inch = 25.4 mm.

TABLE 3.2 Minimum Sizes of Fixture Water Supply Pipes

FIXTURE	MINIMUM PIPE SIZE (inch)
Bathtubs* (60" x 32" and smaller)	½
Bathtubs* (larger than 60" x 32")	½
Bidet	⅜
Combination sink and tray	½
Dishwasher, domestic*	½
Drinking fountain	⅜
Hose bibbs	½
Kitchen sink*	½
Laundry, 1, 2, or 3 compartments*	½
Lavatory	⅜
Shower, single head*	½
Sinks, flushing rim	¾
Sinks, service	½
Urinal, flush tank	½
Urinal, flush valve	¾
Wall hydrant	½
Water closet, flush tank	⅜
Water closet, flush valve	1
Water closet, flushometer tank	⅜
Water closet, one piece*	½

For SI: 1 inch = 25.4 mm, 1 foot = 304.8 mm, 1 pound per square inch = 8.895 kPa.

* *Where the developed length of the distribution line is 60 feet or less, and the available pressure at the meter is a minimum of 35 psi, the minimum size of an individual distribution line supplied from a manifold and installed as part of a parallel water distribution system shall be one nominal tube size smaller than the sizes indicated.*

FIGURE 3.1 Reduced-pressure principal backflow preventer. (*Courtesy of ASPE.*)

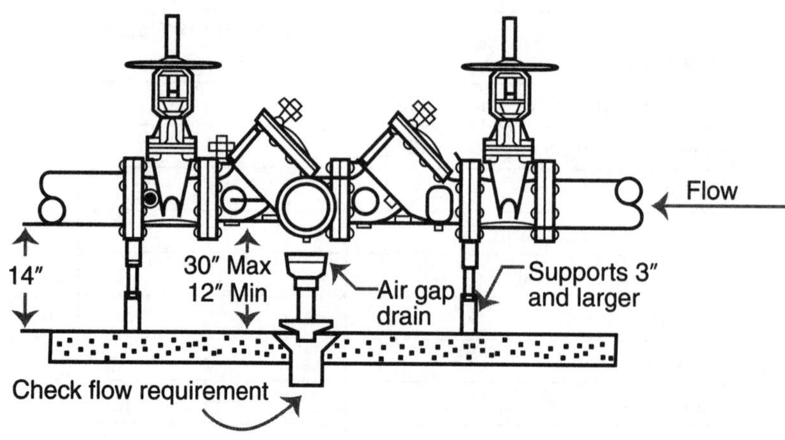

FIGURE 3.2 Double-check-valve backflow preventer. (*Courtesy of ASPE.*)

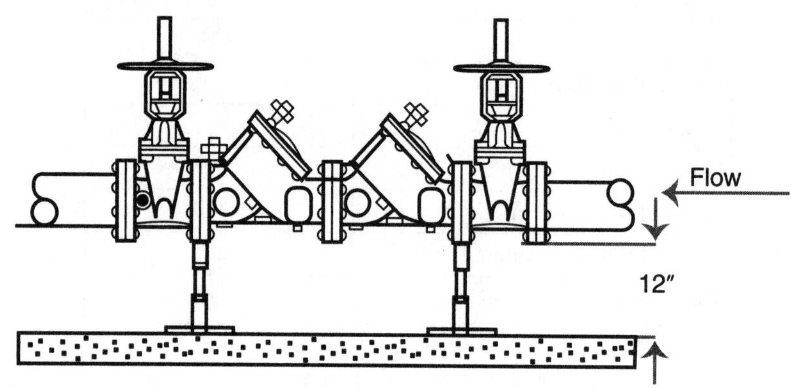

FIGURE 3.3 Atmospheric-type vacuum breaker. (*Courtesy of ASPE.*)

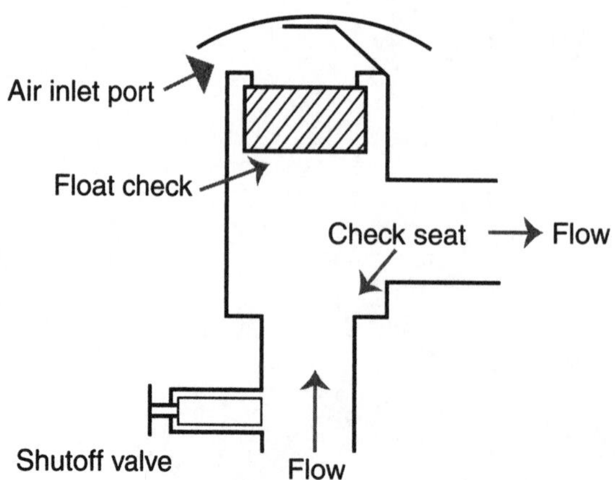

Chapter 4

WATER HEATERS, HEAT EXCHANGERS, AND BOILERS

Hot water has been recognized as a virtual requirement for almost all facilities used by human beings throughout the country. This chapter provides the regulations for installing domestic hot-water heaters. The need to conform to various local codes and standards determines many aspects of the installation of such heaters. The federal government, various other agencies, and local authorities may have specific codes that could be different from those which appear in this text but must be observed. To show the installation of boilers and water heaters, both of which must be installed by plumbing contractors, Figures 4.1 and 4.2 illustrate typical equipment.

One of the most important aspects of modern hot-water heaters is the need to prevent outbreaks of a common disease known as legionnaire's disease. It has been found that water temperatures over 140°F (60°C) will effectively kill the bacteria that causes this disease. Water at this temperature, if it comes into contact with humans, will result in serious scalding. Hot-water heaters are required to heat water to 140°F (60°C). The temperature of the water reaching faucets and showers must be kept at a lower temperature of 110°F (43°C) so that no harm will be done. This is achieved with the use of a mixing valve.

MULTIPLE-CHOICE EXAM

1. Potable water from a tankless water heater shall not exceed which of the following temperatures?

   a. 110°F (43°C) b. 120°F (49°C)

   c. 130°F (54°C) d. 140°F (60°C)

2. Where a potable water heater is used as a space heater, the water temperature for the potable water shall be limited to which of the following temperatures?

   a. 110°F (43°C) b. 120°F (49°C)

   c. 130°F (54°C) d. 140°F (80°C)

3. Where a potable water heater is installed in an attic, how big shall the clear access opening for removal of the water heater be?

   a. 24 × 30 inches (609 × 750 mm)

   b. 20 × 30 inches (500 × 750 mm)

   c. 24 × 24 inches (609 × 609 mm)

   d. 20 × 36 inches (500 × 915 mm)

4. Where a potable water heater is installed in an attic, how big should the clear access on the service side of the water heater be?

   a. 30 × 30 inches (750 × 750 mm)

   b. 20 × 30 inches (500 × 750 mm)

   c. 24 × 24 inches (609 × 609 mm)

   d. 24 × 30 inches (609 × 750 mm)

5. Where a potable water heater is not installed in an attic, how big shall the clear access on the service side of the water heater be?

   a. 24 × 24 inches (609 × 609 mm)

   b. 20 × 30 inches (500 × 750 mm)

   c. 30 × 30 inches (750 × 750 mm)

   d. 24 × 30 inches (609 × 750 mm)

6. When a pan is installed under a water heater, at a minimum, how deep shall the pan be?

   a. 1 inch (25 mm) b. 1½ inches (40 mm)

   c. 2 inches (50 mm) d. 2½ inches (63 mm)

7. At a minimum, the indirect drain from a water heater pan shall be how big?

   a. ⅜ inch (10 DN) b. ½ inch (13 DN)

   c. ¾ inch (20 DN) d. 1 inch (25 DN)

8. What is the requirement for a discharge pipe from a temperature- or pressure-relief valve (or combination):

   a. It must be connected directly to the drainage system.

   b. It need not be trapped.

   c. It must discharge through an air gap.

   d. It may have a valve or tee fitting

9. Mandatory pressures for installed hot-water heaters must be displayed

   a. only where visible.

   b. or the operating pressure must be indicated.

   c. or the tested pressure must be indicated.

   d. or permanently stamped and attached.

10. What is the cold-water line feeding a hot-water heater or tank prohibited from having?

    a. A valve serving other water supplies

    b. Valves that isolate each heater and tank

    c. Valves located adjacent to the heater

    d. Valves within reach of the heater or tank

11. Gas-fired water heaters shall be installed in conformance with which code?

    a. *International Life Safety Code* b. *International Building Code*

    c. *International Plumbing Code* d. *International Fuel Gas Code*

12. Water heaters using solid, liquid, or gas fuel *must not* be installed in a room containing which of the following equipment?

a. Air-handling machinery
b. Electrical equipment
c. Pumps or pumping equipment
d. Mechanical equipment

13. Where earthquakes are a consideration, a hot-water heater shall be installed in accordance with which of the following codes?

a. *International Life Safety Code*
b. *International Building Code*
c. *International Plumbing Code*
d. *International Fuel Gas Code*

14. Which of the following safety devices is often installed on the cold-water supply to a hot-water heater to prevent siphoning?

a. A pressure-reducing valve
b. A reduced-pressure backflow preventer
c. A vacuum-relief valve
d. A check valve

15. What kind of relief valve shall be used on a hot-water heater?

a. A pressure and temperature valve
b. A pressure valve only
c. A temperature valve only
d. A vacuum breaker

16. A pressure-relief valve shall be provided with which of the following?

a. A manual closing mechanism
b. An automatic closing mechanism
c. A way to prevent thermal expansion
d. A shutoff mechanism

17. The outlet of a relief valve must be

a. connected directly to a drain.
b. connected to other discharges.
c. trapped.
d. discharged through an air gap.

18. All hot-water heaters are required to be labeled in what manner?

a. Energy efficiency
b. Accurately
c. Fourth party
d. Third party

19. Where must valve(s) required to isolate the potable water supply on a hot-water heater be located?

a. On the hot-water branch line only

b. On both the hot- and cold-water supplies

c. On the cold-water branch line only

d. On the main water distribution system

20. The potable water supply to a commercial boiler with additives shall be protected with which of the following?

a. A reduced-pressure backflow preventer

b. A double-check-valve backflow preventer

c. A vacuum breaker

d. A shutoff valve

21. The potable water supply to a commercial boiler with no additives shall be protected with which of the following?

a. A double-check-valve backflow preventer

b. A shutoff valve

c. A backflow preventer with an atmospheric vent

d. A vacuum breaker

22. The potable water supply to a residential boiler with no additives shall be protected with which of the following?

a. A submerged relief valve discharge

b. A backflow preventer with an atmospheric vent

c. A vacuum breaker

d. A double-check-valve backflow preventer

23. Heat exchangers used for heat recovery, such as solar heaters, shall protect the potable water supply with which of the following?

a. A backflow preventer with an atmospheric vent

b. A vacuum breaker

c. Double-wall pipe for potable water

d. A double-check-valve backflow preventer

24. Unfired hot-water pressure storage tanks shall have insulation at which of the following ratings?

a. R-10 b. R-12.5

c. R-15 d. R-20.5

25. The potable water discharge routed into a sanitary system from a temperature- and pressure-relief valve from a boiler cannot be drained into which of the following?

a. An open receptor through an air gap

b. An indirect waste receptor

c. A floor drain through an air gap

d. A trapped connection to the sanitary system

26. The drain pan under a water heater is not permitted to drain to which of the following?

a. A direct connection to a sanitary system

b. A floor drain

c. An indirect waste receptor

d. The exterior of the building

27. What is the maximum allowable pressure setting for a temperature- and pressure-relief valve?

a. 125 psi (837 kPa)

b. 150 psi (1,005 kPa)

c. The working pressure of the tank

d. The lesser of the tank working pressure or 150 psi (1,005 kPa)

28. What is the rating of a hot-water heater temperature-relief valve based on?

a. The temperature limit of the heater

b. The heat input of the heater

c. The heat output of the heater

d. The working pressure of the water system

29. How high above the ground surface shall the drain pan under the water heater terminate?

a. 18 inches (450 mm)

b. 34 inches (850 mm)

c. 24 inches (600 mm)

d. 12 inches (300 mm)

30. What is the standard working pressure for a water heater?

a. 150 psi (1,005 kPa)

b. 120 psi (804 kPa)

c. 140 psi (940 kPa)

d. 125 psi (837 kPa)

FIGURE 4.1 Typical domestic hot-water heater.

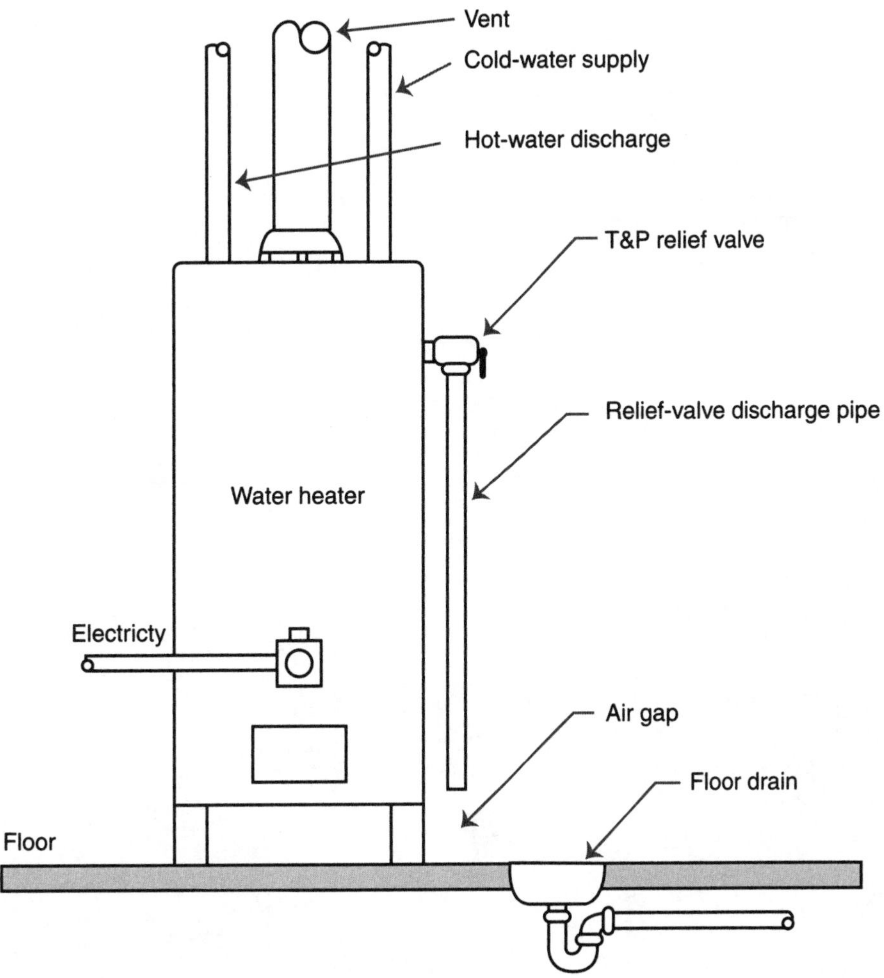

FIGURE 4.2 Typical domestic boiler installation.

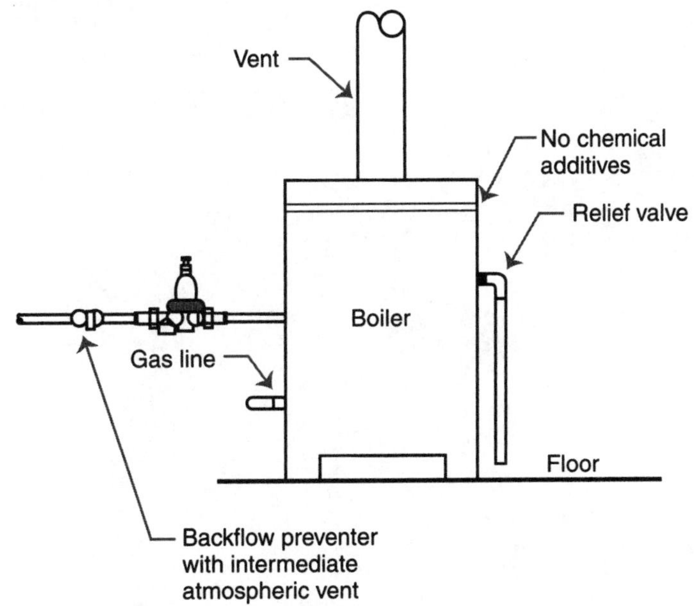

—NOTES—

Chapter 5
SANITARY DRAINAGE

A drainage system is a major component of a facility. The drainage system is sized using the cumulative drainage-fixture-unit (DFU) values of all the connected fixtures into any drainage line and the slope of the drainage line or a vertical stack. The DFU values are provided in the code for all sanitary and vent conditions. A number of tables to determine the DFU value of a fixture and for sizing purposes are provided so that applicants can become familiar with them and be proficient in their use.

1. Table 5.1 provides the DFU values for individual fixtures and trap sizes.
2. Table 5.2 provides DFU values for various trap sizes.
3. Table 5.3 provides the minimum capacity for sewage pumps.
4. Table 5.4 provides the size of horizontal branches and stacks based on DFU.
5. Table 5.5 provides the capacity of building drains and sewers based on DFU.

Optimal performance by drainage piping depends on the size of the pipe and the slope, both of which influence the waste velocity within the drainage system. If the velocity is low and the pipe is oversized, the solids will tend to settle out of suspension and settle to the bottom of the pipe. This eventually will result in drain stoppage. To prevent this from happening, a velocity of 2 feet per second of the effluent has been confirmed by tests over many years.

One of the fundamental requirements of the drainage system is that the drainage piping shall not be reduced in size in the direction of flow.

A cleanout is designed to provide access to the drainage system for the purpose of maintaining a clear opening for the effluent and to provide a means to clear stoppages. The cleanout must be readily accessible for all piping, whether concealed or exposed. Minimum access to these cleanouts is a specific code requirement. It should be extended up to the floor above and flush with that floor to avoid being a tripping hazard. For piping at the ceiling, the cleanout shall be extended to the floor above. A cleanout is often used to connect new fixtures during an alteration, and in such a situation, a new cleanout must be provided elsewhere.

Vents are critical to maintain equilibrium throughout a vertical drainage system. Water falling in drainage stacks creates surges in air pressure and creates a vacuum. For this reason, relief vents that connect the drainage and vent stacks are necessary. The same surges in air pressure and vacuum occur in offsets; therefore, yoke vents are used.

MULTIPLE-CHOICE EXAM

1. Building sewers shall be provided with cleanouts located *not* more than how many feet apart measured from the upstream entrance of that cleanout?

   a. 400 feet (120 m) b. 125 feet (38 m)

   c. 200 feet (60 m) d. 100 feet (30 m)

2. For building sewers that are 8 inches (DN 200) and larger, manholes shall be provided not more than how many feet apart measured from the junction building drain and building sewer?

   a. 200 feet (60 m) b. 100 feet (30 m)

   c. 300 feet (90 m) d. 250 feet (75 m)

3. Cleanouts shall be installed at each change of direction of a building drain or horizontal waste or soil line greater than how many degrees?

   a. 30 degrees b. 15 degrees

   c. 45 degrees d. 22½ degrees

4. Lead bends and traps shall *not* be less than what thickness?

   a. ⅛ inch (3.5 mm) b. ½ inch (13 mm)

   c. 0.187 inch (4.75 m) d. ¼ inch (7 mm)

5. Where more than one change of direction occurs in a run of piping, only one cleanout shall be required for each of how many feet of developed length of the drainage piping?

   a. 45 feet b. 30 feet

   c. 40 feet d . 100 feet

6. Cleanouts shall be the same nominal size as the pipe they serve up to how many inches (mm)?

   a. 6 inches (150 mm) b. 5 inches (125 mm)

   c. 4 inches (100 mm) d. 3 inches (75 mm)

7. For pipes larger than 4 inches (100 mm) nominal size, the minimum size of the cleanout shall be how many inches in diameter?

   a. 4 inches (100 mm) b. 5 inches (125 mm)

   c. 6 inches (150 mm) d. 3 inches (75 mm)

8. Cleanouts on 6-inch (DN 150) and smaller pipes shall be provided with a clearance around the cover of not less than how many inches for rodding?

   a. 12 inches (300 mm) b. 6 inches (150 mm)
   c. 18 inches (450 mm) d. 24 inches (600 mm)

9. The minimum trap size for unlisted fixtures shall be the size of the drainage outlet but not less than how many inches?

   a. 1.25 inches (32 mm) b. 1.5 inches (40 mm)
   c. 2 inches (50 mm) d. 3 inches (75 mm)

10. DFU values for continuous flow into a drainage system shall be computed based on the fact that 1 gpm flow is equivalent to how many fixture units?

   a. 2 b. 3
   c. ½ d. 1

11. Where a stack has a horizontal offset, it is required to have a vent after how many branch intervals?

   a. 3 b. 4
   c. 1 d. 2

12. A building sump pit is *not* permitted to be less than how many inches in diameter?

   a. 24 inches (600 mm) b. 18 inches (460 mm)
   c. 30 inches (750 mm) d. 20 inches (50 mm)

13. A building sump pit is *not* permitted to be less than how many inches in depth?

   a. 18 inches (460 mm) b. 30 inches (750 mm)
   c. 24 inches (610 mm) d. 36 inches (920 mm)

14. The level control shall prevent effluent from rising to within how many inches of the invert of the gravity drain inlet into the sump?

   a. 1 inch (25 mm) b. 2 inches (50 mm)
   c. 3 inches (75 mm) d. 4 inches (100 mm)

15. Pumps shall connect to a building drain a minimum of how many feet from the base of any soil stack, waste stack, or fixture drain?

   a. 7 feet (2.1 m) b. 8 feet (2.4 m)
   c. 9 feet (2.7 m) d. 10 feet (3 m)

16. Ejectors receiving the discharge of water closets shall be capable of handling spherical solids with a diameter of up to and including how many inches?

a. 1 inch (25 mm) b. 2 inches (50 mm)

c. 3 inches (75 mm) d. 1.5 inches (37.5 mm)

17. Grinder pumps or grinder ejectors that receive the discharge of water closets shall have a minimum discharge opening of how many inches?

a. 0.75 inch (19 mm) b. 1 inch (25 mm)

c. 1.25 inches (32 mm) d. 1.5 inches (37 mm)

18. Macerating (grinder) toilet assemblies that serve water closets shall have a minimum discharge opening of how many inches?

a. 0.75 inch (19 mm) b. 1 inch (25 mm)

c. 1.25 inches (32 mm) d. 1.5 inches (37 mm)

19. Where a battery of no more than three sterilizers discharges to an individual receptor, the distance between the receptor and a sterilizer shall not exceed how many feet?

a. 5 feet (1.5 m) b. 6 feet (1.8 m)

c. 7 feet (2.1 m) d. 8 feet (2.4 m)

20. The velocity of airflow in a central vacuum (fluid suction) system shall be less than how many feet per minute?

a. 3,000 feet per minute (900 m/min)

b. 4,000 feet per minute (1,200 m/min)

c. 5,000 feet per minute (1,500 m/min)

d. 6,000 feet per minute (1,800 m/min)

21. The local vent for a bedpan washer shall *not* be less than how many inches in diameter?

a. 1 inch (25 mm) b. 2 inches (50 mm)

c. 2.25 inches (56 mm) d. 2.5 inches (63 mm)

22. Not more than 12 bedpan washers are allowed to be connected to a local vent stack of how many inches?

a. 1 inch (25 mm) b. 2 inches (50 mm)

c. 3 inches (75 mm) d. 4 inches (100 mm)

23. A water supply pipe shall be trapped to form a water seal of no less than how many inches?

a. 1 inch (25 mm)
b. 2 inches (50 mm)
c. 3 inches (75 mm)
d. 4 inches (100 mm)

24. What is the allowed minimum size of a sterilizer vent serving a bedpan steamer?

a. 1 inch (25 mm)
b. 1.5 inches (37 mm)
c. 1.75 inches (44 mm)
d. 2 inches (50 mm)

25. What is the minimum allowed size vent stack serving an instrument sterilizer?

a. 0.75 inch (20 mm)
b. 1 inch (25 mm)
c. 1.5 inches (37 mm)
d. 1.75 inches (44 mm)

26. What is the minimum allowed size of a pressure sterilizer vent stack?

a. 1 inch (25 mm)
b. 1.5 inches (37 mm)
c. 2 inches (50 mm)
d. 2.5 inches (63 mm)

27. What is the minimum permitted slope of horizontal drainage pipe that is 2½ inches (DN 65) in diameter?

a. ¼ inch per foot (2 cm/m)
b. ⅛ inch per foot (1 cm/m)
c. 1⁄16 inch per foot (0.5 cm/m)
d. 1⁄32 inch per foot (0.25 cm/m)

28. A connection of waste water into a drainage system shall have a temperature *not* exceeding which of the following?

a. 110°F (43°C)
b. 120°F (49°C)
c. 130°F (54°C)
d. 140°F (60°C)

29. Back-to-back connections for water closets are permitted when the horizontal distance is greater than how many inches?

a. 24 inches (600 mm)
b. 36 inches (900 mm)
c. 18 inches (450 mm)
d. 30 inches (760 mm)

30. What is the DFU value for a bathtub using Table 5.1?

a. 1 DFU
b. 2 DFUs
c. 3 DFUs
d. 4 DFUs

31. What is the DFU value for a domestic kitchen sink using Table 5.1?

a. 1 DFU
b. 2 DFUs
c. 3 DFUs
d. 4 DFUs

32. What is the DFU value for a private 1.6-gpf water closet using Table 5.1?

a. 1 DFU
b. 2 DFUs
c. 3 DFUs
d. 4 DFUs

33. What is the DFU value for a fixture drain 2 inches (DN 50) in size using Table 5.2?

a. 1 DFU
b. 2 DFUs
c. 3 DFUs
d. 4 DFUs

34. What is the DFU for a horizontal branch 2 inches (DN 50) in size using Table 5.4?

a. 2 DFUs
b. 3 DFUs
c. 4 DFUs
d. 6 DFUs

35. How many DFUs are permitted for a 5-inch (DN 125) building sewer pitched at ⅛ inch per foot (1 cm/m) using Table 5.3?

a. 390 DFUs
b. 480 DFUs
c. 1,000 DFUs
d. 180 DFUs

36. A backwater valve is required to protect plumbing fixtures with flood-level rims below what elevation?

a. Building sewer
b. Exterior septic field tank
c. Building drain
d. Public sewer

37. What is the minimum capacity of a sewage ejector with a discharge pipe that is 2 inches (DN 50) in diameter?

a. 21 gpm (80 L/min)
b. 30 gpm (115 L/min)
c. 46 gpm (175 L/min)
d. 55 gpm (210 L/min)

38. What is the DFU value for a fixture trap 2 inches (DN 50) in size using Table 5.2?

a. 1 DFU
b. 2 DFUs
c. 3 DFUs
d. 4 DFUs

39. When using the *UPC*, at what distance outside the building wall does the designation building drain change to house sewer?

a. At the building wall
b 2 feet (0.6 m)
c. 4 feet (1.2 m)
d. 5 feet (1.5 m)

40. A bathroom group has a 1.6-gpm water closet, one lavatory, and one shower. What is the DFU value using Table 5.1?

a. 4 DFUs
b. 5 DFUs
c. 6 DFUs
c. 7 DFUs

41. What should be the minimum size drain outlet for a central washing facility in a multiple-unit dwelling?

a. 4 inches (200 mm)
b. 2½ inches (63 mm)
c. 3 inches (150 mm)
d. 2 inches (100 mm)

42. What is the maximum number of DFUs allowed into a 6-inch (DN 150) building drain with a slope of ¼ inch per foot using Table 5.5?

a. 575 DFUs
b. 840 DFUs
c. 700 DFUs
d. 1,000 DFUs

43. When conducting a forced sewer test on a building sewer with a pump rated at 20 psi, what is the correct test pressure?

a. 20 psi (45 kPa)
b. 25 psi (167 kPa)
c. 30 psi (200 kPa)
d. 40 psi 9 (45 kPa)

44. When using the *UPC* for horizontal drains within a building, cleanouts shall be located at what maximum intervals?

a. 100 feet (30 m)
b. 75 feet (25 m)
c. 50 feet (15 m)
d. 200 feet (60 m)

45. Drainage connections of a stack containing suds-producing fixtures shall *not* be made within how many feet of a vertical to horizontal change of direction?

a. 2 feet (0.75 m)
b. 4 feet (1.3 m)
c. 6 feet (2 m)
d. 8 feet (2.6 m)

46. When using the *UPC*, a public sewer is considered available if the sewer is located how many feet from a proposed building?

a. 75 feet (25 m)
b. 100 feet (32 m)
c. 200 feet (62 m)
d. 400 feet (125 m)

TABLE 5.1 Drainage Fixture Unit Values for Fixtures and Groups

FIXTURE TYPE	DRAINAGE FIXTURE UNIT VALUE AS LOAD FACTORS	MINIMUM SIZE OF TRAP (inches)
Automatic clothes washers, commercial[a,g]	3	2
Automatic clothes washers, residential[g]	2	2
Bathroom group as defined in Section 202 (1.6 gpf water closet)[f]	5	—
Bathroom group as defined in Section 202 (water closet flushing greater than 1.6 gpf)[f]	6	—
Bathtub[b] (with or without overhead shower or whirpool attachments)	2	$1^1/_2$
Bidet	1	$1^1/_4$
Combination sink and tray	2	$1^1/_2$
Dental lavatory	1	$1^1/_4$
Dental unit or cuspidor	1	$1^1/_4$
Dishwashing machine,[c] domestic	2	$1^1/_2$
Drinking fountain	$^1/_2$	$1^1/_4$
Emergency floor drain	0	2
Floor drains	2	2
Kitchen sink, domestic	2	$1^1/_2$
Kitchen sink, domestic with food waste grinder and/or dishwasher	2	$1^1/_2$
Laundry tray (1 or 2 compartments)	2	$1^1/_2$
Lavatory	1	$1^1/_4$
Shower	2	$1^1/_2$
Service sink	2	$1^1/_2$
Sink	2	$1^1/_2$
Urinal	4	Note d
Urinal, 1 gallon per flush or less	2[e]	Note d
Urinal, nonwater supplied	0.5	Note d
Wash sink (circular or multiple) each set of faucets	2	$1^1/_2$
Water closet, flushometer tank, public or private	4[e]	Note d
Water closet, private (1.6 gpf)	3[e]	Note d
Water closet, private (flushing greater than 1.6 gpf)	4[e]	Note d
Water closet, public (1.6 gpf)	4[e]	Note d
Water closet, public (flushing greater than 1.6 gpf)	6[e]	Note d

For SI: 1 inch = 25.4 mm; 1 gallon = 3.785 L (gpf = gallon per flushing cycle).

a. For traps larger than 3 inches, use Table 709.2.

b. A showerhead over a bathtub or whirlpool bathtub attachment does not increase the DFU value.

c. See Sections 709.2 through 709.4 for methods of computing unit value of fixtures not listed in this table or for rating of devices with intermittent flows.

d. Trap size shall be consistent with the fixture outlet size.

e. For the purpose of computing loads on building drains and sewers, water closets and urinals shall not be rated at a lower DFU value unless the lower values are confirmed by testing.

f. For fixtures added to a dwelling unit bathroom group, add the DFU value of those additional fixtures to the bathroom group fixture count.

g. See Section 406.3 for sizing requirements for fixture drain, branch drain, and drainage stack for an automatic clothes washer standpipe.

TABLE 5.2 DFU Values for Fixture Drains or Traps

FIXTURE DRAIN OR TRAP SIZE (inches)	DRAINAGE FIXTURE UNIT VALUE
$1^1/_4$	1
$1^1/_2$	2
2	3
$2^1/_2$	4
3	5
4	6

For SI: 1 inch = 25.4 mm.

TABLE 5.3 Minimum Capacity of Sewage Pumps or Sewage Ejectors

DIAMETER OF THE DISCHARGE PIPE (inches)	CAPACITY OF PUMP OR EJECTOR (gpm)
2	21
$2^1/_2$	30
3	46

For SI: 1 inch = 25.4 mm; 1 gallon per minute (gpm) = 3.785 L/m.

TABLE 5.4 Horizontal Fixture Branches and Stacks

DIAMETER OF PIPE (inches)	MAXIMUM NUMBER OF DRAINAGE FIXTURE UNITS (dfu)			
		Stacks[b]		
	Total for horizontal branch	Total discharge into one branch interval	Total for stack of three branch intervals or less	Total for stack greater than three branch intervals
$1^1/_2$	3	2	4	8
2	6	6	10	24
$2^1/_2$	12	9	20	42
3	20	20	48	72
4	160	90	240	500
5	360	200	540	1,100
6	620	350	960	1,900
8	1,400	600	2,200	3,600
10	2,500	1,000	3,800	5,600
12	2,900	1,500	6,000	8,400
15	7,000	Note c	Note c	Note c

For SI: 1 inch = 25.4 mm.

a. Does not include branches of the building drain. Refer to Table 710.1(1).

b. Stacks shall be sized based on the total accumulated connected load at each story or branch interval. As the total accumulated connected load decreases, stacks are permitted to be reduce in size. Stack diameters shall not be reduced to less than one-half the diameter of the largest stack size required.

c. Sizing load based on design criteria.

TABLE 5.5 Building Drains and Sewers

DIAMETER OF PIPE (inches)	MAXIMUM NUMBER OF DRAINAGE FIXTURE UNITS CONNECTED TO ANY PORTION OF THE BUILDING DRAIN OR THE BUILDING SEWER, INCLUDING BRANCHES OF THE BUILDING DRAIN[a]			
	Slope per foot			
	$^1/_{16}$ inch	$^1/_8$ inch	$^1/_4$ inch	$^1/_2$ inch
$1^1/_4$	—	—	1	1
$1^1/_2$	—	—	3	3
2	—	—	21	26
$2^1/_2$	—	—	24	31
3	—	36	42	50
4	—	180	216	250
5	—	390	480	575
6	—	700	840	1,000
8	1,400	1,600	1,920	2,300
10	2,500	2,900	3,500	4,200
12	3,900	4,600	5,600	6,700
15	7,000	8,300	10,000	12,000

For SI: 1 inch = 25.4 mm; 1 inch per foot = 83.3 mm/m.

a. The minimum size of any building drain serving a water closet shall be 3 inches.

—NOTES—

Chapter 6

FIXTURES AND FAUCETS

The definition of a plumbing *fixture* is any receptacle or device that is connected to a water supply system and discharges waste, whether clear, cloudy, or sanitary in nature, into a drainage system. A *receptacle* is a fixture that is not directly connected to a water supply but could discharge waste, whether clear or cloudy in nature, into a drainage system. A *faucet* is defined as any device that discharges water with the ability to turn on and off the water supply on demand. This chapter will test the ability of applicants to find a facility's fixture types.

Many plumbers think that fixture choice is up to the property owner for whom the work is being done, and this is generally correct, except for smaller residential projects. The job of choosing proper plumbing fixtures, except for smaller types of projects, is not generally one that a licensed plumber will be required do. Architects typically are responsible for determining the type and minimum fixture requirements for a large building based on occupancy. When the decision is made for the contractor to choose a fixture, a knowledge of the proper code requirements will be necessary. There is no question that the licensed plumber will be required to install all plumbing fixtures.

In residential plumbing, the required fixtures are not difficult to choose. However, in commercial facilities, knowing what types of fixtures are required and in what numbers gets more difficult and is up to the architect or engineer because he or she is required to have the project approved by the building department.

MULTIPLE-CHOICE EXAM

1. What is a receptacle or device called that discharges wastewater, liquid-borne waste materials, or sewage either directly or indirectly to the drainage system?

a. A plumbing appurtenance
b. A plumbing fixture
c. A sanitary connection
d. A sanitary unit

2. What is a receptacle or device called that requires both a water-supply connection and a discharge to a drainage system?

a. A plumbing appurtenance
b. A sanitary unit
c. A plumbing fixture
d. A plumbing conservation device

3. What is a water closet called that is installed in such a way that it does not touch a floor?

a. A carrier water closet
b. A wall-mounted water closet
c. A hanger-hung water closet
d. A wall-hung water closet

4. What is a bathtub called that is equipped and fitted with a circulation piping system, pump, and similar appurtenances and is so designed to accept, circulate, and discharge bathtub water?

a. A spa
b. A bathtub and spa
c. A whirlpool spa
d. A whirlpool bathtub

5. Which of the following locations would *not* be considered to have fixtures intended for public use?

a. Toilet rooms in schools
b. Toilet rooms in gymnasiums
c. Toilet rooms in residences
d. Toilet rooms in hospitals

6. Which of the following locations would *not* be required to have public facilities?

a. Hospitals
b. Parking garages without an attendant
c. Schools
d. Malls

7. Separate facilities are often required for men and women using some types of facilities. Which types are exempt from this requirement?

a. Residential properties
b. Business properties where occupant load total is 15 people or fewer
c. Mercantile occupancies with a maximum occupant load of 100 people or fewer
d. A property that serves food or beverages for consumption within the structure

8. To determine the normal total occupancy load except for specific statistical data, what should be the ratio of males and females?

a. Evenly distributed with 50 percent for each sex
b. Distributed with 75 percent of the fixtures allotted for females
c. Distributed with 75 percent of the fixtures allotted for males
d. Distributed as determined by the owner

9. In most cases, access to toilet facilities for employees must be

a. within a building.
b. accessible from the employees' regular working area.
c. directly into a kitchen.
d. within a distance of 50 feet outside the building.

10. Toilet facilities for employees, in most cases, must be located not more than what distance above or below the employees' regular work area?

a. 50 feet (17m)
b. Two stories
c. 25 feet (9m)
d. One story

11. Toilet facilities for employees, in most cases, must be located in such a way that employees will not have to travel more than how many feet to reach them?

a. 50 feet (17m)
b. 100 feet (34m)
c. 200 feet (66m)
d. 500 feet (168m)

12. What is the minimum width of a water-closet compartment?

a. 30 inches (750 mm) b. 32 inches (800 mm)

c. 36 inches (900 mm) d. 42 inches (1050 mm)

13. What is the minimum compartment depth of a floor mounted water-closet compartment?

a. 30 inches (750 mm) b. 36 inches (900 mm)

c. 60 inches (1,500 mm) d. 48 inches (1,200 mm)

14. What is the minimum distance of clear space required from the center of a water closet to the closest sidewall?

a. 12 inches (300 mm) b. 15 inches (375 mm)

c. 18 inches (450 mm) d. 30 inches (750 mm)

15. What is the maximum temperature of water a shower is allowed to discharge from a individual shower head?

a. 110°F (43°C) b. 115°F (45°C)

c. 120°F (49°C) d. 100°F (43°C)

16. What is the minimum of clear space required from the center of a lavatory to the closest sidewall?

a. 12 inches (300 mm) b. 15 inches (375 mm)

c. 18 inches (450 mm) d. 30 inches (750 mm)

17. When lavatories are installed side by side, what is the minimum distance required between the centerlines of the adjacent fixtures?

a. 12 inches (300 mm) b. 15 inches (375 mm)

c. 18 inches (450 mm) d. 30 inches (750 mm)

18. Fixtures equipped with concealed slip-joint connections, as in the case of a tub waste and overflow, must be provided with a waste outlet of what size?

a. 2 inches (50 mm) b. 2½ inches (63 mm)

c. 1½ inch (38 mm) d. 1¼ inches (32 mm)

19. The water supply lines and fittings for fixtures shall be installed primarily to prevent which of the following situations?

a. Backflow b. Expansion

c. Condensation d. Vibration

20. A waste connection is *not* required for which of the following fixtures?

a. Dishwasher
b. Laundry tray
c. Clothes washer
d. Emergency shower

21. Where are drinking fountains *not* permitted?

a. Restaurants
b. Churches
c. Public rest rooms
d. Malls

22. What is the minimum size of a floor drain?

a. 2 inches (50 mm)
b. 2½ inches (62 mm)
c. 1½ inches (38 mm)
d. 3 inches (75 mm)

23. Water supplies to automatic clothes washers must be protected against backflow by which of the following devices?

a. An air break
b. An air gap
c. A vacuum breaker
d. A check valve

24. Water supplies to garbage grinders and dishwashing machines must be protected against backflow by which of the following?

a. An air break
b. An air gap
c. A vacuum breaker
d. A check valve

25. Which of the following requires an air gap as protection from backflow?

a. Boilers
b. Sinks
c. Bathtubs
d. Dishwashers

26. When designing a men's bathroom in a building classified as assembly, what percentage of urinals can be substituted for water closets?

a. 25 percent
b. 50 percent
c. 67 percent
d. 75 percent

27. Trough urinals are

a. permitted.
b. required.
c. allowed by a code official.
d. not permitted.

28. What is the minimum size vent stack for a pressure instrument sterilizer?

a. 1½ inches (38 mm)
b. 2 inches (50 mm)
c. 2½ inches (62 mm)
d. 3 inches (75 mm)

29. What is the minimum diameter of the strainer that is permitted on the waste side of a shower?

a. 2 inches (50 mm)
b. 2½ inches (62 mm)
c. 3 inches (75 mm)
d. 4 inches (100 mm)

30. Using the *IPC*, what should be the maximum water temperature permitted for a lavatory to deliver?

a. 100°F (39°C)
b. 110°F (43°C)
c. 120°F (49°C)
d. 130°F (54°C)

31. What is the permitted water allowance for low-consumption fixtures?

a. 1.6 gallons per flush (gpf)
b. 1.3 gpf
c. 1.2 gpf
d. 3.6 gpf

32. What is the required outlet pipe size for a floor-mounted back-outlet water closet?

a. 2½ inches (62 mm)
b. 4 inches (100 mm)
c. 5 inches (127 mm)
d. 3 inches (75 mm)

33. When using the *UPC*, how big should the waste outlet for a shower be?

a. 1½ inches (38 mm)
b. 2 inches (50 mm)
c. 2½ inches (62 mm)
d. 3 inches (75 mm)

34. When using the *IPC*, how big should the waste outlet for a shower be?

a. 1½ inches (38 mm)
b. 2 inches (50 mm)
c. 2½ inches (62 mm)
d. 3 inches (75 mm)

The following questions refer to Table 6.1. No unusual conditions are used or to be considered. The tables provided in this chapter are included for illustration purposes only to show the correct use of such tables for your locality. They are not to be used for design purposes and are designed just for taking the test. Caution is to be exercised in the use of this table except for the test given in the area where the test is taken.

35. For a nightclub project, the licensed plumber is asked to determine the number of plumbing fixtures using the A-2 classification. The total expected maximum number of occupants is 300.

a. What will be the number of male water closets required?
b. What will be the number of female water closets required?
c. What will be the total number of lavatories required?
d. How many drinking fountains will be required?

Answers and explanations: The first calculation is to determine the ratio of males to females. The code states that the occupancy would be half male and half female. With 300 people, that would amount to 150 men and 150 women.

a. The answer is a total of 8 (150 ÷ 40 = 3.75, say, 4 for each sex).

b. Same as above.

c. The correct total number of lavatories is 4 (150 ÷ 75 = 2 for each sex).

d. The correct number of drinking fountains is 1 (1 for each 500 people).

36. For a business office building project, the licensed plumber is selected to determine the plumbing fixtures. The total expected maximum number of occupants is 500.

a. What will be the number of male water closets required?

b. What will be the number of female water closets required?

c. What will be the total number of lavatories required?

d. How many drinking fountains will be required?

Answers and explanations: The first calculation is to determine the ratio of males to females. To find the value for each sex, the occupancy should be divided in half. With 500 people, that would amount to 250 men and 250 women. Using the B classification, the answer is a total of 11 water closets for men.

a. The answer is a total of 22 water closets, calculated as follows: For men, the number of water closets is 1 per 25 for the first 50 = 2 required and 1 per 50 for the remainder of the occupancy = 9 (450 ÷ 50 = 9). Thus the total required water closets for men is 9 + 2 = 11.

b. Same as above.

c. The answer is a total of 10 lavatories, calculated as follows: For men, the number of lavatories would be 1 per 40 for the first 80 = 2 and 1 per 80 for the remainder of the occupancy = 2.13, say, 3. Double this for the total number of lavatories required which is 5 + 5 = 10.

d. The answer is a total of 5 drinking fountains (1 for each 100 people).

37. For a high-rise apartment house project, the licensed plumber is asked to determine the number of plumbing fixtures. The total expected number of dwelling units is 300.

a. What will be the number of water closets required?

b. What will be the number of lavatories required?

c. What will be the total number of bathtubs/showers required?

d. What will be the total number of kitchen sinks required?

e. How many clothes washer connections will be required?

Answers and explanations: The first calculation would be to determine the number of dwelling units. This number has been established as 300.

a. Based on 300 dwelling units, the number of water closets required is 300.

b. Based on 300 dwelling units, the number of lavatories required is 300.

c. Based on 300 dwelling units, the number of bathtubs/showers required is 300.

d. Based on 300 dwelling units, the number of kitchen sinks required is 300.

e. The correct number of washer connections answer is 15.

f. The required number of washer connections is 1 per 20 dwelling units. $300 \div 20 = 15$

TABLE 6.1 Minimum Number of Required Plumbing Fixtures

NO.	CLASSIFICATION	OCCUPANCY	DESCRIPTION	WATER CLOSETS (URINALS SEE SECTION 419.2)		LAVATORIES		BATHTUBS/ SHOWERS	DRINKING FOUNTAIN (SEE SECTION 410.1)	OTHER
				MALE	FEMALE	MALE	FEMALE			
1	Assembly	A-1[d]	Theaters and other buildings for the performing arts and motion pictures	1 per 125	1 per 65	1 per 200		—	1 per 500	1 service sink
		A-2[d]	Nightclubs, bars, taverns, dance halls, and buildings for similar purposes	1 per 40	1 per 40	1 per 75		—	1 per 500	1 service sink
			Restaurants, banquet halls, and food courts	1 per 75	1 per 75	1 per 200		—	1 per 500	1 service sink
		A-3[d]	Auditoriums without permanent seating, art galleries, exhibition halls, museums, lecture halls, libraries, arcades, and gymnasiums	1 per 125	1 per 65	1 per 200		—	1 per 500	1 service sink
			Passenger terminals and transportation facilities	1 per 500	1 per 500	1 per 750		—	1 per 1000	1 service sink
			Places of worship and other religious services	1 per 150	1 per 75	1 per 200		—	1 per 1000	1 service sink
		A-4[d]	Coliseums, arenas, skating rinks, pools, and tennis courts for indoor sporting events and activities	1 per 75 for the first 1500 and 1 per 120 for the remainder exceeding 1500	1 per 40 for the first 500 and 1 per 60 for the remainder exceeding 1500	1 per 200	1 per 150	—	1 per 1000	1 service sink
		A-5	Stadiums, amusement parks, bleachers, and grandstands for outdoor sporting events and activities	1 per 75 for the first 1500 and 1 per 120 for the remainder exceeding 1500	1 per 40 for the first 1500 and 1 per 60 for the remainder exceeding 1500	1 per 200	1 per 150	—	1 per 1000	1 service sink
2	Business	B	Buildings for the transaction of business, professional services, other services involving merchandise, office buildings, banks, light industrial and similar uses	1 per 25 for the first 50 and 1 per 50 for the remainder exceeding 50		1 per 40 for the first 80 and 1 per 80 for the remainder exceeding 50		—	1 per 100	1 service sink
3	Educational	E	Educational facilities	1 per 50		1 per 50		—	1 per 100	1 service sink
4	Factory and industrial	F-1 and F-2	Structures in which occupants are engaged in work fabricating, assembly, or processing of products or materials	1 per 100		1 per 100		See Section 411	1 per 400	1 service sink

(continued)

TABLE 6.1 Minimum Number of Required Plumbing Fixtures *(Continued)*

NO.	CLASSIFICATION	OCCUPANCY	DESCRIPTION	WATER CLOSETS (URINALS SEE SECTION 419.2)		LAVATORIES		BATHTUBS/ SHOWERS	DRINKING FOUNTAIN (SEE SECTION 410.1)	OTHER
				MALE	FEMALE	MALE	FEMALE			
5	Institutional	I-1	Residential care	1 per 10		1 per 50		1 per 8	1 per 100	1 service sink
		I-2	Hospitals, ambulatory nursing home patients[b]	1 per room[c]		1 per room[c]		1 per 15	1 per 100	1 service sink per floor
			Employees, other than residential care[b]	1 per 25		1 per 35		—	1 per 100	—
			Visitors, other than residential care	1 per 75		1 per 100		—	1 per 500	—
		I-3	Prisons[b]	1 per cell		1 per cell		1 per 15	1 per 100	1 service sink
			Reformatories, detention centers, and correctional centers[b]	1 per 15		1 per 15		1 per 15	1 per 100	1 service sink
		I-4	Adult day care and child care	1 per 15		—		—	1 per 100	1 service sink
6	Mercantile	M	Retail stores, service stations, shops, salesrooms, markets, and shopping centers	1 per 500		1 per 750		—	1 per 1000	1 service sink
7	Residential	R-1	Hotels, motels, boarding houses (transient)	1 per sleeping unit		1 per sleeping unit		1 per sleeping unit	—	1 service sink
		R-2	Dormitories, fraternities, sororities, and boarding houses (not transient)	1 per 10		1 per 10		1 per 8	1 per 100	1 service sink
			Apartment house	1 per dwelling unit		1 per dwelling unit		1 per dwelling unit	—	1 kitchen sink per dwelling unit; 1 automatic clothes washer connection per 20 dwelling units
		R-3	One- and two-family dwellings	1 per dwelling unit		1 per dwelling unit		1 per dwelling unit	—	1 kitchen sink per dwelling unit; 1 automatic clothes washer connection per dwelling unit
		R-4	Residential care/ assisted living facilities	1 per 10		1 per 10		1 per 8	1 per 100	1 service sink

(continued)

TABLE 6.1 Minimum Number of Required Plumbing Fixtures *(Continued)*

NO.	CLASSIFICATION	OCCUPANCY	DESCRIPTION	WATER CLOSETS (URINALS SEE SECTION 419.2)		LAVATORIES		BATHTUBS/ SHOWERS	DRINKING FOUNTAIN (SEE SECTION 410.1)	OTHER
				MALE	FEMALE	MALE	FEMALE			
8	Storage	S-1, S-2	Structures for the storage of goods, warehouses, storehouse and freight depots. Low and Moderate Hazard.	1 per 100		1 per 100		See Section 411	1 per 1000	1 service sink

a. *The fixtures shown are based on one fixture being the minimum required for the number of persons indicated or any fraction of the number of persons in indicated. The number of occupants shall be determined by the* International Building Code.
b. *Toilet facilities for employees shall be separate from facilities for inmates or patients.*
c. *A single-occupant toilet room with one water closet and one lavatory serving not more than two adjacent patient sleeping units shall be permitted where such room is provided with direct access from each patient room and with provisions for privacy.*
d. *The occupant load for seasonal outdoor seating and entertainment areas shall be included when determining the minimum number of facilities required.*

—NOTES—

Chapter 7

INDIRECT WASTE

The importance of preventing any backup or reverse flow from a contaminated waste source from entering or being drawn into the potable water supply or sanitary system cannot be overemphasized. An indirect waste connection to a drainage system provides a nonmechanical means of completely protecting the public from the effects of any liquid reversing flow from non-potable water systems, appliances, and equipment that serve some special purpose. Another aspect of the indirect waste provisions is to prevent contamination of the potable water supply from anything that poses any hazard at all to the public, as determined by the code official.

To clarify the distinction between an air gap and an air break, an *air gap* provides the highest level of protection and prevents any possible cross connection between a potable water supply and any drainage system. An air gap is created when the indirect waste discharge pipe terminates above the flood level of the waste receptor. An *air break* is created when the indirect waste discharge pipe extends below the flood level of the waste receptor but terminates above the trap seal of the receptor or receiving fixture. It is used when there is no concern about the backup of waste but where splashing of the discharge is to be avoided. The difference between an air gap and air break is illustrated in Figure 7.1.

MULTIPLE-CHOICE EXAM

1. For the *IPC*, what shall the air-gap distance be between the outlet of an indirect waste pipe and the receptor?
   a. 2 inches (50 mm)
   b. The distance of the waste opening
   c. Twice the distance of the waste opening
   d. 1 inch (25 mm)

2. For the *UPC*, what shall the air-gap distance be between the outlet of an indirect waste pipe and the receptor?
   a. 2 inches (50 mm)
   b. 1½ inches (38 mm)
   c. 3 inches (75 mm)
   d. 1 inch (25 mm)

3. A vent, if required, from indirect waste piping shall conform to which of the following?
   a. It is permitted to combine with other vents.
   b. It need not extend to the outside air.
   c. If the pipe is 15 feet (5 m) long, a vent is required.
   d. If the pipe is 15 feet (5 m) long, a vent is not required.

4. For the *IPC*, how long should the total developed length of an indirect waste pipe discharging into a waste receptor be?

a. Not longer than 2 feet (0.6 m)

b. Not longer than 3 feet (1 m)

c. Not longer than 4 feet (1.2 m)

d. Not longer than 4 feet, 6 inches (1.4 m)

5. For the *UPC*, how long should the length of an indirect waste pipe discharging into a receptor be?

a. Not longer than 5 feet (1.5 m)

b. Not longer than 4 feet (1.2m)

c. Not longer than 3 feet (1 m)

d. No limit

6. What type of fixtures are permitted to be drained by an indirect waste pipe?

a. Sterilizers

b. Relief-valve discharge

c. Clear potable water discharge

d. Discharge from a kitchen sink

7. Equipment and fixtures used for what purpose shall not discharge through an indirect waste pipe?

a. Food handling

b. Food preparation

c. Kitchen sink

d. Food storage

8. A domestic washing machine is *not* permitted to discharge

a. through an air gap.

b. past the trap of a kitchen sink.

c. through an air break.

d. through a Y-branch on the sink tailpiece.

9. A standpipe serving a clothes washer discharging into a waste receptor shall

a. extend no more than 30 inches (750 mm) above the trap.

b. extend less than 30 inches (750 mm) above the trap.

c. have a trap 6 inches (150 mm) above the floor.

d. be installed below the floor.

10. Waste receptors for indirect wastes shall be

a. installed in a bathroom.

b. installed in a laundry room.

c. covered with a removable strainer.

d. installed in a ventilated space.

11. A standpipe serving as a receptor shall *not* be permitted to

a. extend a minimum of 18 inches (450 mm) above the trap.

b. extend a minimum 48 inches above trap (450 mm) above the trap.

c. be individually trapped.

d. connect to a sanitary drain with no trap.

12. What equipment is not required to discharge into the drainage system through an indirect waste pipe?

a. Food-handling equipment

b. Sterilizers

c. A surgeon's sink

d. Distillation equipment

13. Indirect waste requirements do not pertain to which of the following?

a. Refrigeration condensate

b. Toilet facilities

c. Boiler blow-off

d. General-area floor drains

14. If a floor drain is located within an area subject to freezing, the waste line serving the drain must

a. discharge directly to the sanitary system.

b. be trapped.

c. discharge inside the area.

d. discharge outside the area.

15. The direct connection to the sanitary drainage system of what fixture may be permitted?

a. A floor drain with a backflow preventer

b. A dishwasher

c. A pool water filter backwash discharge

d. A pot sink

16. What is an approved fixture or device called that is permitted to accept the discharge from indirect waste pipes?

a. An air-gap

b. A relief valve

c. An air-break

d. A receptor

17. When using the *UPC*, indirect waste piping exceeding 5 feet (1.5 m) in length is *not* permitted to have which of the following?

a. A trap if it is less than 15 feet (4.5 m) long

b. A trap if it exceeds 15 feet (4.5 m) in length

c. A trap if it is no longer than 5 feet (1.5 m)

d. Discharge into a receptor with no trap

18. When using the *UPC*, what type of appliance is *not* considered to be a plumbing fixture and may not be permitted to be drained by an indirect waste pipe?

a. The tailpiece of a lavatory

b. A receptor outlet

c. A mechanical room drip

d. A pump drip

19. Where are indirect waste receptors *not* permitted?

a. In boiler rooms

b. In ventilated spaces

c. In bathrooms

d. In soiled-linen utility rooms

20. The drain from a waste receptor shall have what installed on the outlet?

a. A vent only

b. A backflow preventer

c. A check valve

d. A trap with a vent

21. The waste-water discharge from what swimming pool equipment is *not* permitted to connect indirectly to a drainage system through an air gap?

a. Pool deck drains

b. The swimming pool main drain

c. Pool area showers

d. Filter backwash

22. What is the difference between an air break and an air gap?

a. Both end above a receptor's flood level.

b. An air break ends above a receptor's flood level.

c. An air gap ends above a receptor's flood level.

d. Both end below a receptor's flood level.

23. What determines the size of a general indirect waste receptor?

a. The number of indirect waste pipes

b. That needed to prevent splashing

c. Minimum discharge into the receptor

d. The area where installed

24. For the *UPC*, what is the minimum size of an indirect waste pipe?

a. 1 inch (25 mm)
b. 1¼ inches (31 mm)
c. ¾ inch (19 mm)
d. 2 inches (50 mm)

25. For the *UPC*, how long should the indirect waste pipe from a sterilizer be?

a. Not longer than 20 feet
b. Not longer than 10 feet
c. Not longer than 15 feet
d. Not longer than 5 feet

26. For the *UPC*, indirect waste drip pipes shorter than 15 feet in length in no case should be less than what size?

a. ¾ inch (19 mm)
b. ½ inch (13 mm)
c. 1 inch (25 mm)
d. 1¼ inches (31 mm)

27. Water of what temperature is not permitted to enter a plumbing system through an indirect waste pipe?

a. 120°F (49°C)
b. 125°F (51°C)
c. 130°F (54°C)
d. 140°F (60°C)

28. Condensate waste drains from air-conditioning equipment should *not* slope less than which of the following?

a. 1⁄16 inch per foot (0.5 cm/m)
b. ⅛ inch per foot (1.0 cm/m)
c. ¼ inch per foot (2.0 cm/m)
d. ½ inch per foot (0.4 cm/m)

29. For the *UPC*, an air-conditioning condensate drain is permitted to connect indirectly into the drainage system through which of the following?

a. A dry well
b. A leach pit
c. The tailpiece of a plumbing fixture
d. It is permitted to connect into the drainage system directly.

30. For the *UPC*, sterilizers should be drained indirectly by a waste pipe into a receptor no longer than which of the following?

a. 5 feet (1.5 m)
b. 10 feet (3 m)
c. 15 feet (4.5 m)
d. 20 feet (6 m)

FIGURE 7.1 Typical indirect waste discharge.

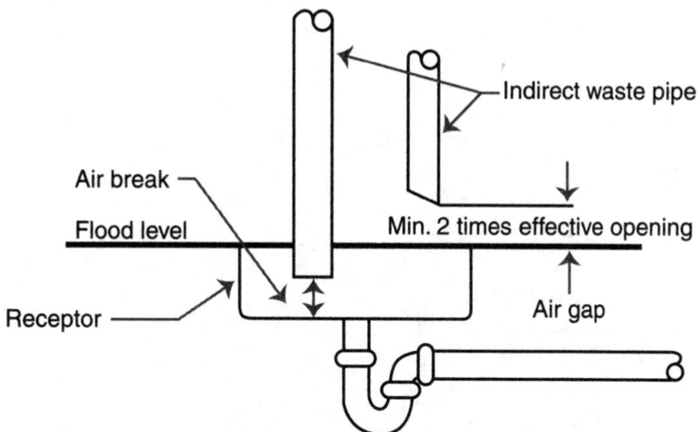

—NOTES—

Chapter 8
VENTS

The purpose of a vent system in a building is to equalize the air-pressure fluctuations that occur in a drainage system from flowing water and to limit it to 1 inch (25 mm) of water column. The two primary reasons this is necessary is to prevent the loss of fixture traps from partial vacuums and to allow the smooth flow of water in the drainage system.

The trap of a fixture prevents sewer gas, odors, and other dangerous gases from entering a habitable space. Without proper venting, a trap seal could be backsiphoned down the drain, leaving the building unprotected from sewer gas, odors, and airborne disease-causing organisms. When a trap is not vented properly, there is a risk that the water that forms the seal will be siphoned down the drain, leaving the trap without enough water to block odors and gases. The trap may be replenished when a small quantity of water is run through it, but the next time a large volume of water is drained, the trap could be emptied again because of negative air pressure. This, of course, is not an acceptable risk. Therefore, traps shall be vented to avoid this problem. For example, if you were to fill a bathtub with water and then open the drain, the water would pass through the trap and down the drain. Without additional air (such as from a vacuum breaker), the volume of water draining into the pipe could create a siphonic action that might pull the trap seal from adjacent fixtures. Should this happen, fixtures would be left unprotected from the release of odors and gases from the drainage system.

A drainage system will not work very well without vents. Vents provide air circulation within a drainage system. They allow the drains to flow faster, and they prevent siphonic action in the traps. Fixtures that are not vented adequately drain slowly. Another problem is presented when the end of a flexible hose is submerged in a contaminated source and a vacuum is created when there is a break in the supply line. Once the siphon in a hose has been started, the flow of contaminated water will continue in the hose until air is introduced into the hose or until the hose is raised above the level of the liquid being siphoned. The air is introduced by a vacuum breaker.

Not all vertical vents are dry and extended to open air. Some are wet vents, meaning that the vent for one fixture serves as a drain for another fixture that has no vent. In combination waste-and-vent systems, the drains are oversized to allow air circulation in the oversized drain pipe above the flow of drainage. There are many options for venting the upper portions of a plumbing system. It is also possible to use air-admittance valves for venting purposes.

It will be easier to gain knowledge of vents if you have a good understanding of why the vent system is sized based on the total DFUs that are connected to the vent and the developed length of the vent. This information is given in Table 8.1. In order to give you such an understanding of the basic concepts, Figures 8.1 and 8.2 illustrate the details of the various arrangements and explain the elements that make up much of a vent system.

Single-stack venting systems and air-admittance valves are not permitted in some localities. Reading of the local code will show whether questions relating to such systems will be of value to the applicant. If those vents are not permitted, skip the questions.

MULTIPLE-CHOICE EXAM

1. Which of the following vent systems must be independent of a sanitary vent system?

a. Chemical vent system
b. Continuous vent system
c. Wet vent system
d. Circuit vent

2. What is a vent that connects to one or more individual vents with a stack vent or a vent stack known as?

a. A yoke vent
b. A wet vent
c. A main vent
d. A branch vent

3. What is a primary vent in a venting system where five vent branches are connected called?

a. A main vent
b. A stack vent
c. A yoke vent
d. A vent stack

4. What is a vent that serves only one fixture drain called?

a. A fixture branch
b. A circuit vent
c. An individual vent
d. A common vent

5. When a soil or waste stack is extended above the highest horizontal drain, which of the following is created?

a. A yoke vent
b. A vent stack
c. A relief vent
d. A stack vent

6. A vent connection at the base of a stack should connect

a. to the building drain.
b. above the lowest waste branch.
c. 10 diameters past the stack.
d. upstream of the lowest waste branch.

7. What are vents intended primarily to protect?

a. Traps
b. Trap seals
c. Fixture outlets
d. The vent stack

8. What is the minimum size of any individual vent pipe?

a. Not less than 1 inch (DN 25)
b. Not less than 1½ inches (DN 32)
c. Not less than 2½ inches (DN 63)
d. Not less than 2 inches (DN 50)

9. The diameter of a vent stack must be, at a minimum, equal to which of the following?
   a. The size of the drain being served
   b. 1¼ inches (31 mm)
   c. Twice the size of the drain being served
   d. 2 inches (50 mm)

10. When determining the developed length of a vent, measure from the furthest point of the vent connection to the drainage system to the point of which of the following?
   a. The connection to a vent stack
   b. The connection to a stack vent
   c. Termination into open air
   d. The building drain

11. How must a vent system for chemical waste terminate or run?
   a. Not run horizontally
   b. Connect to a vent stack
   c. Connect to a stack vent
   d. Through the roof to open air

12. Individual vents must be connected to what point?
   a. Crown of the trap
   b. Bottom of the trap
   c. Fixture drain
   d. Fixture branch

13. The size of any vent with a developed length of more than 40 feet (34 m) must be increased by how much?
   a. One pipe size
   b. Two pipe sizes
   c. 50 percent
   d. No increase is required.

14. The length of a pipeline that is measured along the centerline of the pipe and fittings is called what?
   a. The combined length
   b. The determined length
   c. The developed length
   d. The maximum length

15. What is required when a soil or waste stack consists of 10 branch intervals or more?
   a. A vent stack
   b. A stack vent
   c. A combined vent
   d. A relief vent

16. Every vent stack must terminate where?
   a. Through a roof
   b. Outdoors
   c. Near a window
   d. 10 feet (3 m) from a window

17. Vents must protect trap seals in such a way that the seal is not subjected to a pneumatic pressure differential equal to or exceeding what?

a. A 1-inch (25 mm) water column

b. A $1\frac{1}{2}$-inch (38 mm) water column

c. A 2-inch (50 mm) water column

d. A $2\frac{1}{2}$ inch (73 mm) water column

18. When a roof is used for some purpose other than weather protection, the extension of a vent penetrating the roof must *not* be less than what length?

a. 6 inches (1.50 mm)
b. 12 inches (300 mm)
c. 4 feet (1.3 m)
d. 7 feet (2.3 m)

19. Vent terminals must *not* be installed directly beneath which of the following?

a. Windows
b. Vent extension
c. Ventilating openings
d. Doors

20. What must all vent stacks connect to?

a. The base of a drainage stack
b. The base of a sewer
c. The base of a fixture
d. The base of a waste stack

21. Vent terminals may not be installed where?

a. In attics
b. Through walls
c. Through roofs
d. Under an overhang of a building

22. When a vent pipe is installed on the outside of a building where freezing temperatures may occur, the vent must be protected from freezing by which of the following methods?

a. Placed adjacent to a door
b. Heat
c. Placed adjacent to a window
d. Placed adjacent to a hot pipe

23. Vents that terminate through a wall must be at least what dimension, in feet, from the lot line of the adjacent property?

a. 5 feet (1.5 m)
b. 15 feet (4.5 m)
c. 10 feet (3 m)
d. 20 feet (6 m)

24. When a vent stack is connected to a building drain, the connection must be made within how many diameters downstream of the drainage stack?

a. 2 diameters
b. 4 diameters
c. 5 diameters
d. 10 diameters

25. Vents that terminate through a wall must be at least how many feet above the ground.

a. 5 feet (1.5 m)
b. 10 feet (3 m)
c. 15 feet (4.5 m)
d. 20 feet (6 m)

26. All dry vents must rise vertically to a point at least how many inches above the flood-level rim of the highest trap or trapped fixture being vented?

a. 4 inches (100 mm)
b. 6 inches (150 mm)
c. 12 inches (48 mm)
d. 42 inches (1050 mm)

27. All vents should be graded and connected in a manner that will allow them to do what?

a. Drain back to a drainage pipe by gravity
b. Have a grade of ½ inch per foot
c. Maintain a grade of ⅛ inch per foot
d. Drain below the centerline of the pipe

28. A vent connection to what drainage fixture is allowed below the weir of a trap?

a. Lavatory
b. Kitchen sink
c. Water closet
d. Laundry tray

29. How many traps are permitted to be vented by a common trap?

a. 1
b. 2
c. 3
d. 4

30. Wet venting is permitted for what combination of fixtures?

a. Two bathroom groups on the same floor
b. Two bathroom groups on adjacent floors
c. No limit on same floor
d. Three bathroom groups on the same floor

31. If five or more branch intervals are located above a horizontal offset of a drainage stack, the offsets must be

a. no more than 22½ degrees.
b. trapped.
c. no more than 45 degrees.
d. vented by a yoke vent.

32. Sheet lead for vent-pipe flashings used in field-constructed flashings should weigh not less than how many pounds per square foot?

a. 2 pounds per square foot (10 kg/m^2)

b. 3 pounds per square foot (15 kg/m^2)

c. 5 pounds per square foot (25 kg/m^2)

d. 8 pounds per square foot (24 kg/m^2)

33. When sheet copper is used as a flashing material for vents, it must have a weight of *not* less than which of the following?

a. 8 ounces per square foot

b. 6 ounces per square foot

c. 10 ounces per square foot

d. 12 ounces per square foot

34. When island vents are used, they should be installed with which of the following?

a. A size that is half the drain line

b. Trap adapters

c. Sanitary tees

d. Cleanouts

35. Island vents must rise vertically to what point of the fixture being vented?

a. Above the flood-level rim of the fixture

b. Above the drainage outlet

c. At least 6 inches (150 mm) above the flood-level rim

d. 12 inches (300 mm) above base cabinet

36. When a stack vent is provided for a waste stack, how must the stack vent be sized?

a. One pipe diameter smaller than the waste stack

b. No less than 4 inches (100 mm) in diameter

c. The same size as the waste stack

d. One pipe size larger than the waste stack

37. When there is a danger of frost closure, the size increase shall take place at least how many inches below the roof or inside wall where the vent penetrates?

a. 6 inches (150 mm)

b. 18 inches (450 mm)

c. 12 inches (300 mm)

d. 8 inches (200 mm)

38. A combination waste and vent system is not permitted to receive waste from where?

a. Floor drains

b. Sinks

c. Water closets

d. Drinking fountains

39. A wet vent system is not permitted to receive waste from where?

a. Water closet
b. Lavatory
c. Shower
d. Emergency floor drain

40. When a vent pipe is connected to a vent stack or a stack vent, the connection must be made at least how many inches above the flood-level rim of the highest fixture being vented?

a. 4 inches (100 mm)
b. 6 inches (150 mm)
c. 8 inches (200 mm)
d. 42 inches (1050 mm)

41. Vents that are roughed in for future fixtures are not permitted to do which of the following?

a. Connect to the vent system
b. Be labeled as a vent
c. Be not less than half the size of the drain served
d. Be the size of the drain served

42. The maximum slope allowed on a horizontal combination of waste and vent shall *not* exceed which of the following values?

a. 2 percent slope
b. 4 percent slope
c. 8 percent slope
d. 12 percent slope

43. A relief vent must be provided for every drainage stack that has more than how many branch intervals?

a. 10
b. 5
c. 8
d. 4

44. Island-fixture venting is allowed for which of the following fixtures?

a. Urinals
b. Bidets
c. Sinks
d. Water closets

45. Which of the following fixtures is *not* permitted to be served by a combination waste-and-vent system?

a. Floor drains
b. Sinks
c. Lavatories
d. Water closets

46. A waste stack of more than 10 branch intervals must be provided with a relief vent from the top floor every

a. 15 stories.
b. 12 stories.
c. 10 stories.
d. 8 stories.

47. A fixture has a 3-inch (150-mm) trap, a 3-inch (150-mm) drain, and a pipe slope of ⅛ inch per foot (1 cm/m). How far from the vent may the trap for the fixture be placed?

a. 6 feet (1.8 m) b. 8 feet (2.4 m)

c. 10 feet (3 m) d. 12 feet (3.6 m)

48. A fixture has a 2-inch (50-mm) trap, a 2-inch (50-mm) drain, and a pipe slope of ¼ inch per foot (2 cm/m). How far from the vent may the trap for the fixture be placed?

a. 42 inches (1,050 mm) b. 8 feet (2.4 m)

c. 6 feet (1.8 m) d. 4 feet (1.2 m)

49. A single stack system is *not* permitted to serve more than how many water closets?

a. 2 water closets b. 3 water closets

c. 4 water closets d. Unlimited water closets

50. For a single-stack system, one water closet should discharge into a 3-inch (75-mm) horizontal branch within what developed length?

a. 12 inches (300 mm) b. 18 inches (450 mm)

b. 24 inches (600 mm) c. 30 inches (750 mm)

51. For a single-stack system, water closets are *not* permitted to connect to a stack greater than what dimension?

a. 2 feet (0.6 m) b. 3 feet (1 m)

c. 4 feet (1.2 m) d. 5 feet (1.5 m)

52. For a single-stack system, the maximum vertical drop from a urinal shall be how many inches?

a. 2 inches (50 mm) b. 3 inches (75 mm)

c. 4 inches (100 mm) d. 6 inches (150 mm)

53. For a single-stack system, what is the maximum length for fixtures other than water closets?

a. 12 feet (3.6 m) b. 10 feet (3 m)

c. 8 feet (2.4 m) d. 5 feet (1.5 m)

54. For a single-stack system, for stacks higher than two branch intervals, what is the distance that stacks are *not* permitted to receive discharge?

a. The lower three floors b. The lower two floors

c. The lowest floor d. No restriction

55. When using an air-admittance valve, where is such a value permitted to be installed?

a. Only for fixtures on two adjacent floors

b. Only for fixtures on adjacent floors

c. Not permitted

d. Only for fixtures on the same floor

56. When using an air-admittance valve, what minimum dimension is their installation permitted to be above a horizontal drain being vented?

a. 10 inches (250 mm)
b. 8 inches (200 mm)
c. 4 inches (100 mm)
d. 6 inches (150 mm)

57. When using an air-admittance valve, how many inches above insulation materials is the valve permitted to be installed?

a. 6 inches (150 mm)
b. 8 inches (200 mm)
c. 4 inches (100 mm)
d. 10 inches (250 mm)

58. What minimum size vent is permitted for a pressure sterilizer?

a. 1½ inches (37 mm)
b. 2 inches (50 mm)
c. 2½ inches (63 mm)
d. 3 inches (75 mm)

59. When using *UPC*, approved pipe of what material may be used underground?

a. Galvanized steel
b. Stainless steel pipe
c. Transite pipe
d. Copper pipe

60. Each vent from a fixture should rise vertically above the flood level of a fixture to what elevation before offsetting horizontally?

a. 4 inches (100 mm)
b. 6 inches (150 mm)
c. 8 inches (200 mm)
d. 12 inches (300 mm)

61. Vents that extend through a roof should terminate at least what distance above the roof?

a. 6 inches (150 mm)
b. 4 inches (100 mm)
c. 8 inches (200 mm)
d. 12 inches (300 mm)

62. When using the *IPC*, to prevent frost closure of a vent terminal through a roof, what is the minimum size permitted?

a. 6 inches (150 mm)
b. 4 inches (100 mm)
c. 3 inches (75 mm)
d. 2 inches (50 mm)

63. When using the *UPC*, to prevent frost closure of a vent terminal through a roof, what is the minimum size permitted?

a. 6 inches (150 mm)
b. 4 inches (100 mm)
c. 3 inches (75 mm)
d. 2 inches (50 mm)

64. When using the *IPC*, what minimum dimension for a yoke vent connection to a vent stack above floor level is permitted?

a. 36 inches (900 mm)
b. 42 inches (1,050 mm)
c. 50 inches (1,250 mm)
d. 72 inches (1,800 mm)

TABLE 8.1 Size and Developed Length of Stack Vents and Vent Stacks

DIAMETER OF SOIL OR WASTE STACK (inches)	TOTAL FIXTURE UNITS BEING VENTED (DFU)	MAXIMUM DEVELOPED LENGTH OF VENT (feet)[a] DIAMETER OF VENT (inches)										
		1¼	1½	2	2½	3	4	5	6	8	10	12
1¼	2	30										
1½	8	50	150	—	—	—	—	—	—	—	—	—
1½	10	30	100									
2	12		75	200								
2	20	30	50	150		—	—	—	—	—	—	—
2½	42	26	30	100	300							
3	10		42	150	360	1,040						
3	21	—	32	110	270	810	—	—	—	—	—	—
3	53		27	94	230	680						
3	102			86	210	620						
4	43	—	25	35	85	250	980	—	—	—	—	—
4	140			27	65	200	750					
4	320			23	55	170	640					
5	540	—	—	21	50	150	580		—	—	—	—
5	190				28	82	320	990				
5	490				21	63	250	760				
5	940	—	—	—	18	53	210	670	—	—	—	—
5	1,400				16	49	190	590				
6	500					33	130	400	1,000			
6	1,100	—	—	—	—	26	100	310	780	—	—	—
6	2,000					22	84	260	660			
6	2,900					20	77	240	600			
8	1,800	—	—	—	—		31	95	240	940	—	—
8	3,400						24	73	190	729		
8	5,600						20	62	160	610		
8	7,600	—	—	—	—	—	18	56	140	560		—
10	4,000							31	78	310	960	
10	7,200							24	60	240	740	
10	11,000	—	—	—	—	—	—	20	51	200	630	—
10	15,000							18	46	180	571	
12	7,300								31	120	380	940
12	13,000	—	—	—	—	—	—	—	24	94	300	720
12	20,000								20	79	250	610
12	26,000								18	72	230	500
15	15,000	—	—	—	—	—	—	—		40	130	310
15	25,000									31	96	240
15	38,000	—	—	—	—	—	—	—	—	26	81	200
15	50,000									24	74	180

For SI: 1 inch = 25.4 mm; 1 foot = 304.8 mm.

[a] *The developed length shall be measured from the vent connection to the open air.*

FIGURE 8.1 Vent detail sheet. (*Courtesy M. Frankel.*)

Branch vent. A branch vent is a vent that connects one or more individual or common vents to a vent stack or a stack vent.

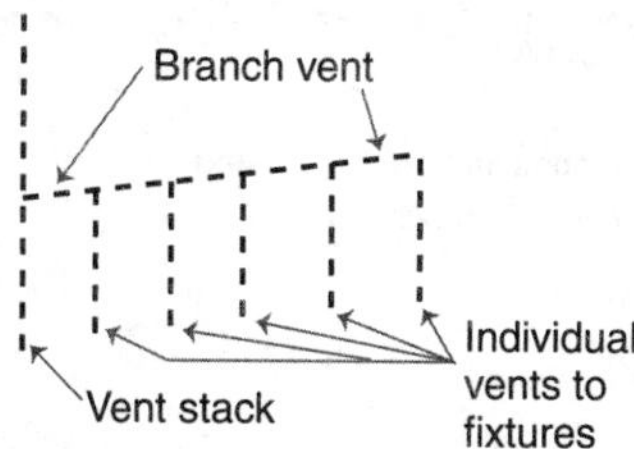

Common vent. A common vent connects two fixtures, with the single vent line serving both.

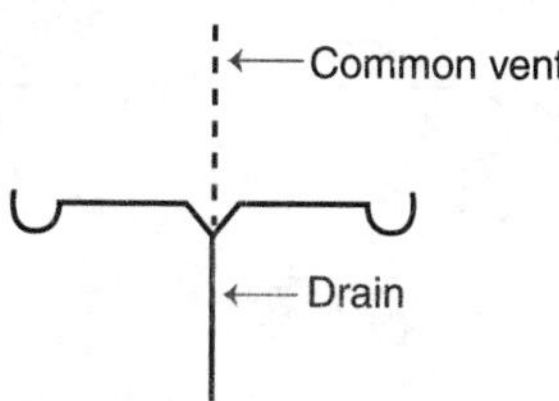

Continuous vent. A continuous vent is a vertical vent that is a continuation of the waste to which it is connected.

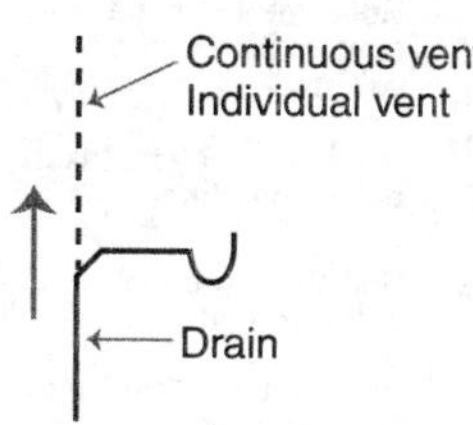

Circuit vent. A circuit vent is a branch vent that serves two or more traps and extends from in front of the last fixture connection to a vent stack.

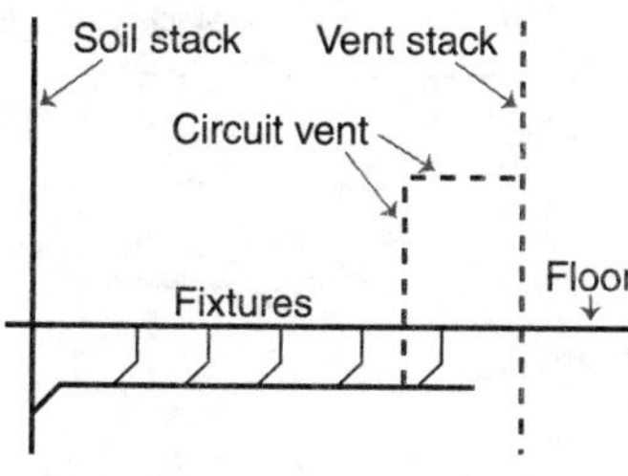

Loop vent. A loop vent is a branch vent that serves two or more traps and extends from in front of the last fixture connection to a stack vent.

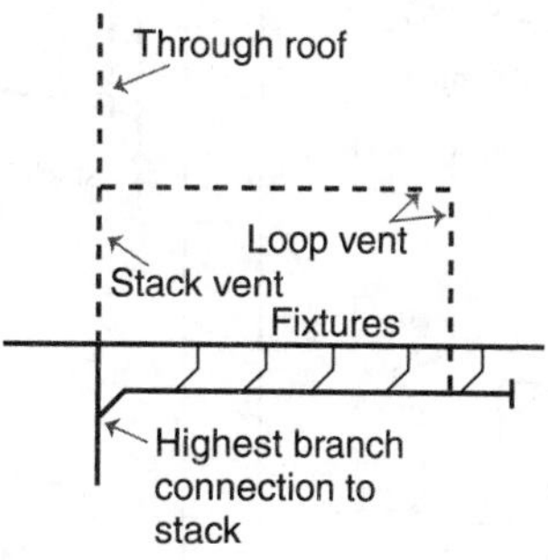

Vent stack. A vertical pipe extending one or more stories and terminating in the outside air allowing circulation of air into and out of a drainage system.

Stack vent. A stack vent is the extension of a soil or waste stack above the highest horizontal connection to that stack.

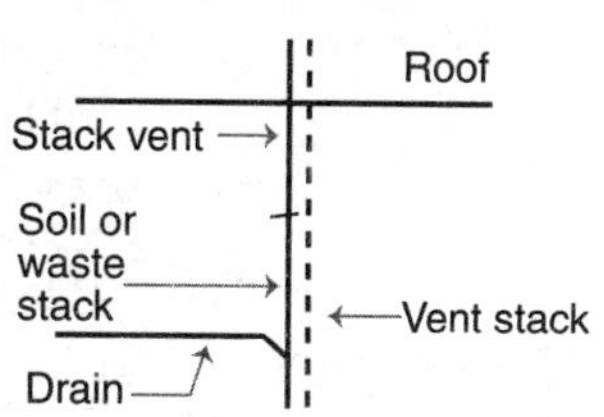

Wet vent. A vent into which a fixture other than a water closet can discharge.

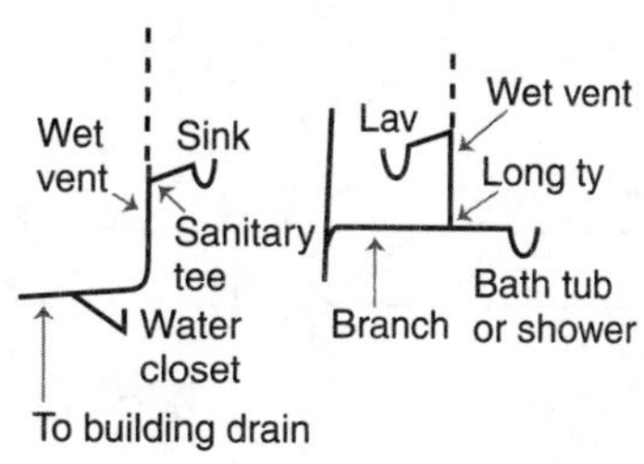

Branch interval.

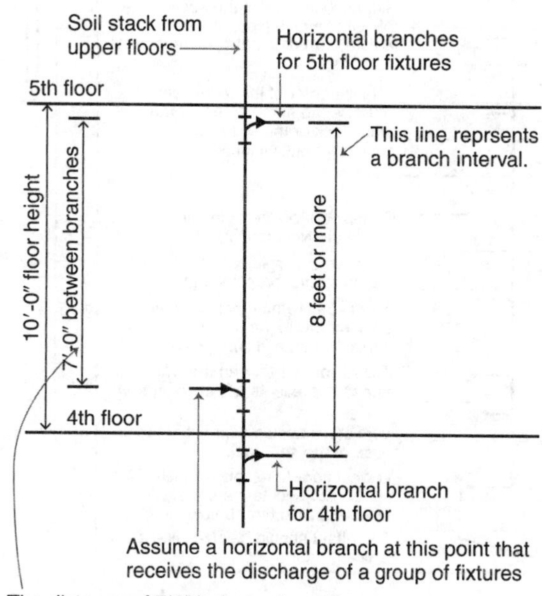

FIGURE 8.2 Vent detail sheet.

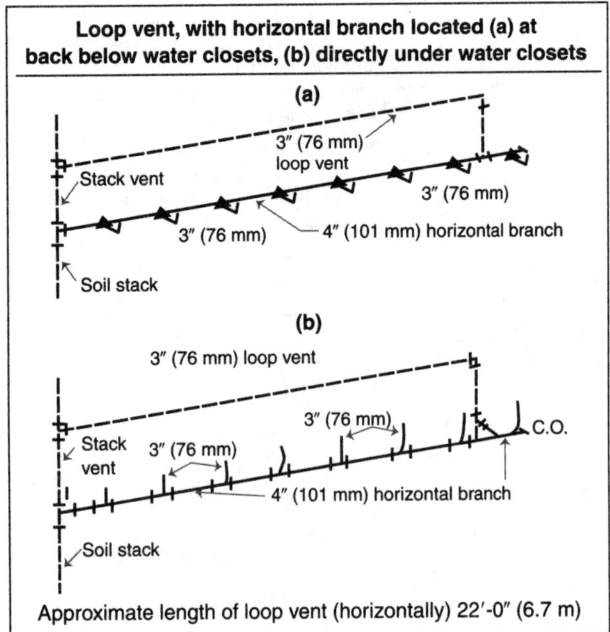

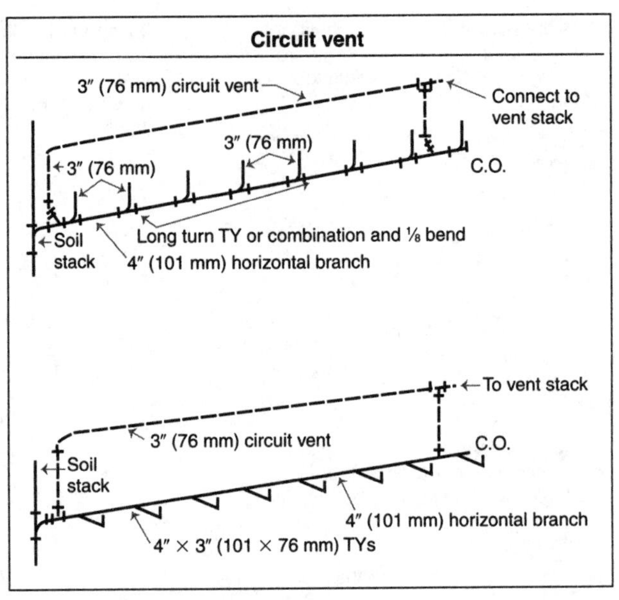

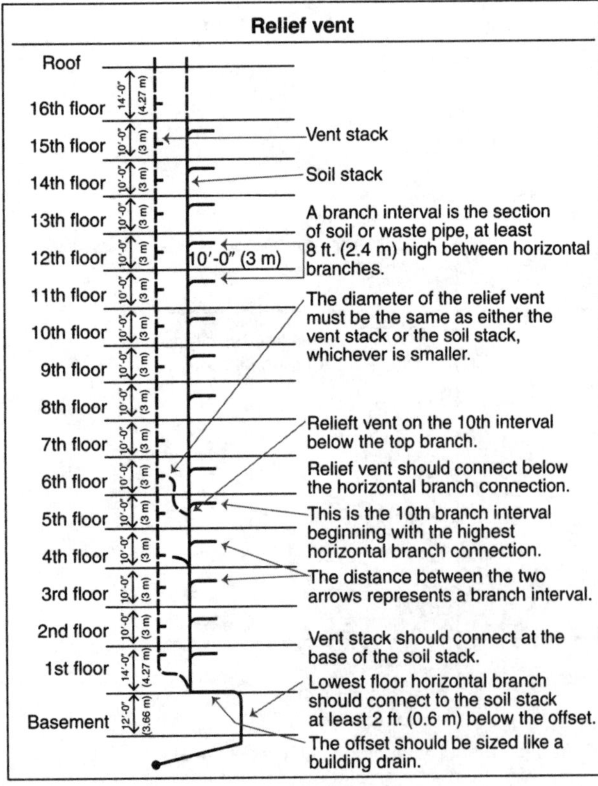

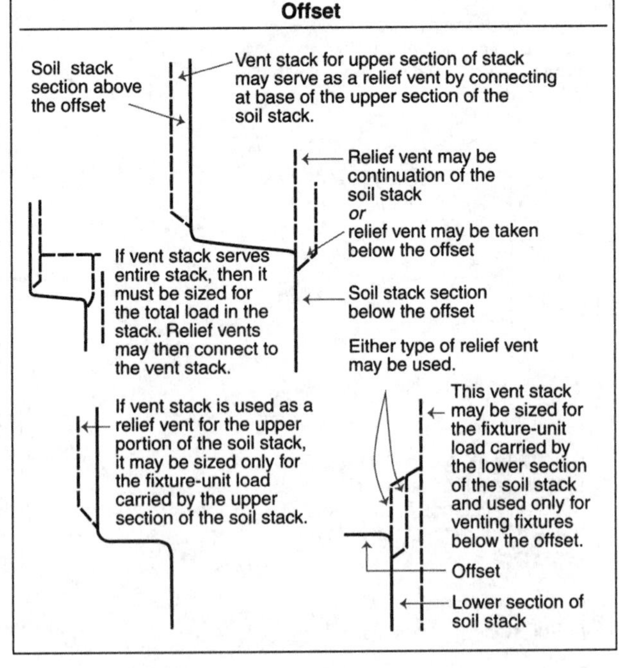

Chapter 9
STORM WATER DRAINAGE

A storm water drainage system is easy to understand. Its basic purpose is to collect precipitation (rain and snow) from roofs and other adjacent areas of the site at the same rate that it collects there, including subsoil drainage piping, and discharge this runoff to an approved point of disposal. When an engineer, owing to code or design, wishes to alleviate overcapacity of a sewer, a controlled roof drain may be installed to delay the heaviest volume of runoff for a longer time. Another important purpose is to prevent the structural failure of flat roofs owing to excessive buildup and runoff of rainwater.

To calculate the piping size for a storm water drainage system, it is necessary to determine the amount of rainfall per hour in the area where the facility is located, the area on which rain will fall, and the pitch of the conveyance pipe. Then correct use of the approved tables in the code will give the size based on allowable roof area, pitch, and shape of the pipe. The horizontal and vertical piping is sized using separate tables in the code. It is not permitted to reduce the pipe size in the direction of flow. It also is not permitted to reduce the pipe size from horizontal to vertical. To size roof drains, determine the rainfall rate and the square feet of roof area. Using the table for circular vertical conductors and leaders, find the size of the drain by using the square-foot figure in the table equal to or greater than the calculated roof area.

As a protective measure to prevent possible structural failure of a roof, a secondary roof drainage system that will allow water to drain from the roof under the most extreme circumstances must be provided. The secondary drainage system is not the same for the *IPC* and the *UPC.* For the *IPC,* the secondary system consists of a completely separate secondary roof drain or open pipe installed at a higher elevation than the primary drain (to prevent debris from accumulating) that discharges to a point on the site that would be visible to the building occupants or a scupper (an opening in the side of the parapet) that allows water to run down the side of a building under all conditions. This is illustrated diagrammatically in Figure 9.1. For the *UPC,* it is permitted, but not mandated, for the secondary drainage system to connect to the primary system discharge to the site above grade. The secondary piping shall be connected to the vertical portion of the primary system. This is illustrated diagrammatically in Figure 9.2. As a licensed plumber, you will have to know how to install storm water systems. The job of system design and sizing is mostly left to engineers and architects, but the licensing exam will test your knowledge of the subject.

The installation of a drainage system to convey storm water is not the same as that of a system meant to transfer sewage and waste water. The hydraulic properties of flowing water are the same for both storm water and sanitary flow. Many of the principles are the same, but still there are distinct differences. Some locations allow the commingling of storm water and waste water, but most jurisdictions prohibit the combined use of such a sewer system.

In talking about a sanitary drainage system, a vertical pipe that conveys waste from one story to another is called a *stack.* If you are installing a drainage system for storm water, a similar pipe is called a *stack,* a *leader,* or a *conductor* depending on the area of the country where it is located.

Subsoil drains are yet another difference when dealing with storm water. These are drains that collect subsurface water and convey it to a point of disposal. *Area drains* are another example of how storm water differs from sanitary drainage. An area drain is similar in terms of plumbing to a floor drain, but it collects storm water and connects to the storm water sewer.

If you were given a detailed set of blueprints for a commercial building and asked to label all the drains as either sanitary or storm drains, the task could be very difficult if you do not understand the differences between the two types of drainage. What appears to be a floor drain in an open-area outside stairway actually will be an area drain. The point of discharge would be the determining factor. This type of confusion could cost you valuable points on the scoring of the licensing exam.

MULTIPLE-CHOICE EXAM

1. Except for one- and two-family dwellings where permitted, storm water shall drain where?

a. Into a separate storm water sewer

b. Into a private sewage disposal system

c. Into a separate sanitary system

d. Into a separate chemical waste system

2. What is the minimum diameter of a sump pit for the discharge of storm water?

a. 36 inches (900 mm)
b. 30 inches (750 mm)
c. 24 inches (576 mm)
d. 18 inches (450 mm)

3. What is a separate building drain called that conveys only storm water but no sewage?

a. A sanitary drain
b. A surface drain
c. A storm sewer
d. A combined drain

4. What is a drain called that collects water from below grade and conveys it to a point of disposal?

a. A combined drain
b. A gray-water drain
c. An underground drain
d. A subdrain

5. Which of the following fittings is *not* permitted for installation in a storm drainage system?

a. A 45-degree bend
b. A 90-degree bend
c. A cleanout
d. A blind plug

6. Which of the following commonly carries both storm water and sewage?

a. A combined sewer
b. A sanitary sewer
c. A mixed drain
d. A subsoil drain

7. Which of the following is *not* permitted to be used in a storm water drainage system?

a. Drainage fittings that are commonly used for sanitary drainage

b. Schedule 40 plastic fittings

c. Fittings that retard the flow of storm water

d. Cast-iron fittings

8. What are pipes used to convey storm water from inside a building to a point of disposal called?

a. Stacks
b. Wet stacks
c Leaders
d. Conductors

9. Except as locally approved, the size of vertical pipes used to convey storm water is based on what criterion?

a. A 50-year storm
b. A 25-year storm
c. A 100-year storm
d. A 10-year storm

10. Where are roof drains inside a building designed to discharge storm water?

a. Into leaders
b. Into conductors
c. Into storm drains
d. Into the sanitary system

11. Under normal conditions, horizontal drains for storm water should have what minimum pitch?

a. $\frac{1}{16}$ inch per foot (0.5 cm/m)
b. $\frac{1}{8}$ inch per foot (1 cm/m)
c. $\frac{1}{4}$ inch per foot (2 cm/m)
d. $\frac{1}{2}$ inch per foot (4 cm/m)

12. When a building subdrain is installed below the level of a public sewer, the subdrain must discharge where?

a. Into a septic tank
b. Into a sump pit
c. Into a receiving tank
d. Into the sanitary system

13. How deep should a sump pit for the discharge of storm water be?

a. A minimum of 36 inches (900 mm)
b. A minimum of 30 inches (750 mm)
c. A minimum of 24 inches (600 mm)
d. A minimum of 18 inches (450 mm)

14. Conductors shall not be used as which of the following?

a. Gray-water pipes
b. Waste pipes
c. Vent pipes
d. Soil pipes

15. When a subsoil drain is subject to backflow, it must be protected with which of the following?

a. A vacuum breaker
b. A gate valve
c. A backwater valve
d. A backflow preventer

16. Subsoil drains may *not* discharge into which of the following?

a. A trapped area drain
b. A sump pit
c. A dry well
d. An ejector pit

17. A controlled-flow system design is based on which of the following?

a. The size of the sanitary sewer
b. The size of the storm sewer
c. The local rainfall rate
d. The size of the local conductor

18. What is a drainage opening through a parapet wall called?

a. An overflow scupper
b. A roof gutter
c. A roof drain
d. A drain opening

19. The pitch of horizontal storm drainage piping shall *not* be less than which of the following?

a. 2 percent
b. 1 percent
c. ½ percent
d. ¼ percent

20. The vertical walls that direct water onto a roof shall be calculated at which of the following rates?

a. No consideration
b. Full area added
c. ¼ area added
d. ½ area added

21. When is a secondary roof drainage required?

a. When the roof perimeter will contain water
b. Under all circumstances
c. When the primary system is undersized
d. Only when connected to a sanitary system

22. If a secondary roof drainage system is required, where shall the discharge occur?

a. Into the sanitary system
b. Into the primary storm water system
c. Above grade in a visible area
d. Into a sump pit inside the building

23. When using the *IPC*, where a continuous discharge from a sump pit discharges into a storm drain or sewer, how should the flow rate be calculated based on 1 inch of rainfall per hour?

a. 1 gpm for each 150 square feet (3.8 L for each 124 m^2)
b. 1 gpm for each 125 square feet (3.8 L for each 103 m^2)
c. 1 gpm for each 75 square feet (3.8 L for each 63 m^2)
d. 1 gpm for each 96 square feet (3.8 L for each 80 m^2)

24. Where a subsoil drain is installed, such as that below footings, where is the discharge *not* permitted to flow?

a. Into a trapped area drain
b. Into the sanitary system
c. Into a dry well
d Into a sump pit

25. Who should design controlled-flow roof drainage system?

a. A professional engineer
b. A licensed plumber
c. The owner of the property
d. The superintendent of construction

26. When using the *UPC*, where a continuous discharge from a sump pit discharges into a storm drain or sewer, how shall the flow rate be calculated based on 4 inches of rainfall per hour?

a. 1 gpm for each 150 square feet (3.8 L for each 124 m^2)
b. 1 gpm for each 24 square feet (3.8 L for each 103 m^2)

The following questions will test your understanding of the tables and figures present in the code. Refer to Tables 9.1, 9.2, 9.3, and 9.4 and Figures 9.3, 9.4, and 9.5. No unusual conditions exist or are to be considered. These tables are included for illustration purposes only. They are invalid for all purposes except to show the correct use of tables expected to be included in the test for your locality. They are not to be used when taking the actual test and are not for design purposes. The necessary information will be given to you at the test.

Examples of sizing

27. A commercial office building contract has been awarded, and you will be required to install the storm water work. The roof is 8,000 square feet in area with four roof drains, and the location is New York City. The pitch of drainage piping is ¼ inch per foot. Find the following:

a. What is the rainfall rate?
b. What is the size of the roof drains?
c. After all the drains are connected, what will be the final size of the horizontal pipe?
d. What will be the size of the vertical leader?
e. What will be the size of the horizontal building sewer?

Calculations

a. Using Figure 9.3, you will find that the New York area has an average rainfall rate of 3.0 inches per hour.

b. 8,000 square feet ÷ 4 = 2,000 square feet. Using Table 9.3, circular vertical leaders, 3 inches per hour of rainfall and 2,000 square feet, the size of each drain would be 3 inches.

c. With 8,000 square feet, a pitch of ¼ inch per foot, and an average rainfall of 3 inches per hour, using Table 9.1, the size of the horizontal pipe from all four roof drains would be 6 inches.

d. With 8,000 square feet, using Table 9.4, for circular vertical leaders, the size would be 5 inches. Since the code will not permit a reduction in size, the correct size of the vertical pipe would be 6 inches.

e. Using Table 9.1 with an average of 3 inches rainfall, 8,000 square feet, and ¼-inch pitch, the horizontal building sewer would be 6 inches in size.

TABLE 9.1 Size of Horizontal Storm Drainage Piping

SIZE OF HORIZONTAL PIPING (Inches)	HORIZONTALLY PROJECTED ROOF AREA (Square Feet) RAINFALL RATE (Inches per Hour)					
	1	2	3	4	5	6
	⅛ unit vertical in 12 units horizontal (1 percent slope)					
3	3,288	1,644	1,096	822	657	548
4	7,520	3.760	2,506	1,800	1,504	1.253
5	13,360	6,680	4,453	3,340	2,672	2,227
6	21,400	10,700	7,133	5,350	4,280	3,566
8	46,000	23,000	15,330	11,500	9,200	7,600
10	82,800	41,400	27,600	20,700	16,580	13,800
12	133,200	66,600	44,400	33,300	26,650	22,200
15	218,00	109,00	72,800	59,500	47,600	39,650
	¼ unit vertical in 12 units horizontal (2 percent slope)					
3	4,640	2.320	1,546	1,160	928	773
4	10,600	5,300	3.533	2,650	2,210	1,766
5	18,880	9,440	6,293	4,720	3,776	3,146
6	30,200	15,100	10.066	7,550	6,040	5,033
8	65,200	32,600	21,733	16,300	13,040	10,866
10	116,800	58,400	38,950	29,200	23,350	19,450
12	188,000	94,000	62,600	47,000	37,600	31,350
15	336,000	168,000	112,000	84,000	67,250	56,000
	½ unit vertical in 12 units horizontal (4 percent slope)					
3	6,576	3,288	2,295	1,644	1,310	1,096
4	15,040	7,520	5,010	3,760	3,010	2,500
5	26,720	13,360	8,900	6,680	5,320	4,450
6	42,800	21,400	13,700	10,700	8,580	7,140
8	92,000	46,000	30,650	23,000	18,400	15,320
10	171,600	85,800	55,200	41,400	33,150	27,600
12	266,400	133,200	88,800	66,600	53,200	44,400
15	476,000	238,000	158,800	119,000	95,300	79,250

For SI: 1 inch = 25.4 mm; 1 square foot = 0.0929 m².

TABLE 9.2 Size of Semicircular Roof Gutters

DIAMETER OF GUTTERS (Inches)	HORIZONTALLY PROJECTED ROOF AREA (Square Feet)					
	RAINFALL RATE (Inches per Hour)					
	1	2	3	4	5	6
	1/16 unit vertical in 12 units horizontal (0.5 percent slope)					
3	680	340	226	170	136	113
4	1,440	720	480	360	288	240
5	2,500	1,250	834	625	500	416
6	3,840	1,920	1,280	960	768	640
7	5,520	2,760	1,840	1,380	1,100	918
8	7,960	3,980	2,655	1,990	1,590	1,325
10	14,400	7,200	4,800	3,600	2,880	2,400
	1/8 unit vertical in 12 units horizontal (1 percent slope)					
3	960	480	320	240	192	160
4	2,040	1,020	681	510	408	340
5	3,520	1,760	1,172	880	704	587
6	5,440	2,720	1,815	1,360	1,085	905
7	7,800	3,900	2,600	1,950	1,560	1,300
8	11,200	5,600	3,740	2,800	2,240	1,870
10	20,400	10,200	6,800	5,100	4,080	3,400
	1/4 unit vertical in 12 units horizontal (2 percent slope)					
3	1,360	680	454	340	272	226
4	2,880	1,440	960	720	576	480
5	5,000	2,500	1,668	1,250	1,000	834
6	7,680	3,840	2,560	1,920	1,536	1,280
7	11,040	5,520	3,860	2,760	2,205	1,840
8	15,920	7,960	5,310	3,980	3,180	2,655
10	28,800	14,400	9,600	7,200	5,750	4,800
	1/2 unit vertical in 12 units horizontal (4 percent slope)					
3	1,920	960	640	480	384	320
4	4,080	2,040	1,360	1,020	816	680
5	7,080	3,540	2,360	1,770	1,415	1,180
6	11,080	5,540	3,695	2,770	2,220	1,850
7	15,600	7,800	5,200	3,900	3,120	2,600
8	22,400	11,200	7,460	5,600	4,480	3,730
10	40,000	20,000	13,330	10,000	8,000	6,660

For SI: 1 inch = 25.4 mm; 1 square foot = 0.0929 m².

TABLE 9.3 Size of Circular Vertical Conductors and Leaders

DIAMETER OF LEADER (Inches)[a]	HORIZONTALLY PROJECTED ROOF AREA (Square Feet) RAINFALL RATE (Inches per Hour)											
	1	2	3	4	5	6	7	8	9	10	11	12
2	2,880	1,440	960	720	575	480	410	360	320	290	260	240
3	8,800	4,400	2,930	2,200	1,760	1,470	1,260	1,100	980	880	800	730
4	18,400	9,200	6,130	4,600	3,680	3,070	2,630	2,300	2,045	1,840	1,675	1,530
5	34,600	17,300	11,530	8,650	6,920	5,765	4,945	4,325	3,845	3,460	3,145	2,880
6	54,000	27,000	17,995	13,500	10,800	9,000	7,715	6,750	6,000	5,400	4,910	4,500
8	116,000	58,000	38,660	29,000	23,200	19,315	16,570	14,500	12,890	11,600	10,545	9,600

For SI: 1 inch = 25.4 mm; 1 square foot = 0.0929 m².

[a] *Sizes indicated are the diameter of circular piping. This table is applicable to piping of other shapes, provided that the cross-sectional shape fully encloses a circle of the diameter indicated in this table. For rectangular leaders, see code. Interpolation is permitted for pipe sizes that fall between those listed in this table.*

TABLE 9.4 Size of Rectangular Vertical Conductors and Leaders

DIMENSIONS OF COMMON LEADER SIZES Width x Length (Inches)[a, b]	HORIZONTALLY PROJECTED ROOF AREA (Square Feet) RAINFALL RATE (Inches per Hour)											
	1	2	3	4	5	6	7	8	9	10	11	12
1¾ × 2½	3,410	1,700	1,130	850	680	560	480	420	370	340	310	280
2 × 3	5,540	2,770	1,840	1,380	1,100	920	790	690	610	550	500	460
2¾ × 4¼	12,830	6,410	4,270	3,200	2,560	2,130	1,830	1,600	1,420	1,280	1,160	1,060
3 × 4	13,210	6,600	4,400	3,300	2,640	2,200	1,880	1,650	1,460	1,320	1,200	1,100
3½ × 4	15,900	7,950	5,300	3,970	3,180	2,650	2,270	1,980	1,760	1,590	1,440	1,320
3½ × 5	21,310	10,650	7,100	5,320	4,260	3,550	3,040	2,660	2,360	2,130	1,930	1,770
3¾ × 4¾	21,960	10,980	7,320	5,490	4,390	3,660	3,130	2,740	2,440	2,190	1,990	1,830
3¾ × 5¼	25,520	12,760	8,500	6,380	5,100	4,250	3,640	3,190	2,830	2,550	2,320	2,120
3½ × 6	27,790	13,890	9,260	6,940	5,550	4,630	3,970	3,470	3,080	2,770	2,520	2,310
4 × 6	32,980	16,490	10,990	8,240	6,590	5,490	4,710	4,120	3,660	3,290	2,990	2,740
5½ × 5½	44,300	22,150	14,760	11,070	8,860	7,380	6,320	5,530	4,920	4,430	4,020	3,690
7½ × 7½	100,500	50,250	33,500	25,120	20,100	16,750	14,350	12,560	11,160	10,050	9,130	8,370

For SI: 1 inch = 25.4 mm; 1 square foot = 0.0929 m².

[a] *Sizes indicated are nominal width × length of the opening for rectangular piping.*

[b] *For shapes not included in this table, Equation 9.1 should be used to determine the equivalent circular diameter* D_e *of rectangular piping for use in interpolation using the data from code.*

$D_e = (\text{width} \times \text{length})^{1/2}$ (Eq. 9.1)

where D_e *= equivalent circular diameter;* D_e*, width, and length are in inches.*

FIGURE 9.1 Typical *IPC* roof drain arrangement. (*Courtesy of ASPE.*)

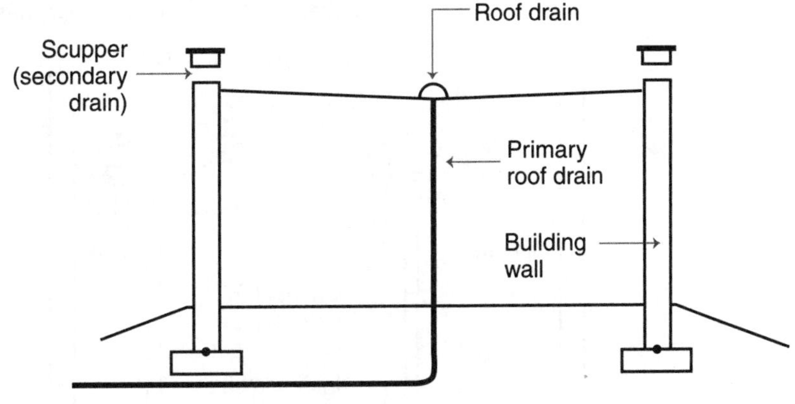

Primary roof drain only

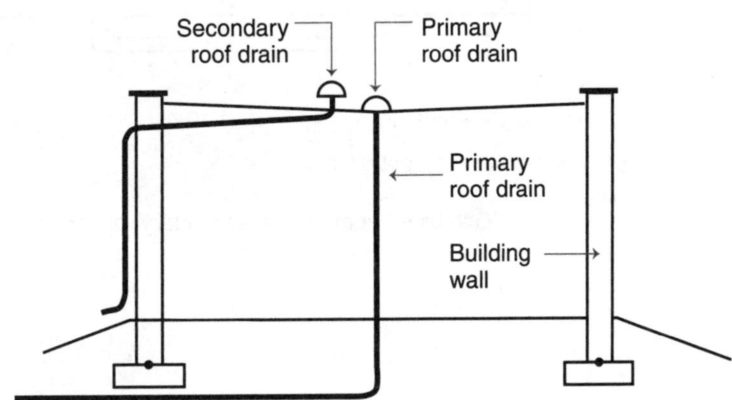

Separate primary and secondary roof drains

FIGURE 9.2 Typical *UPC* roof drain arrangement. (*Courtesy of ASPE.*)

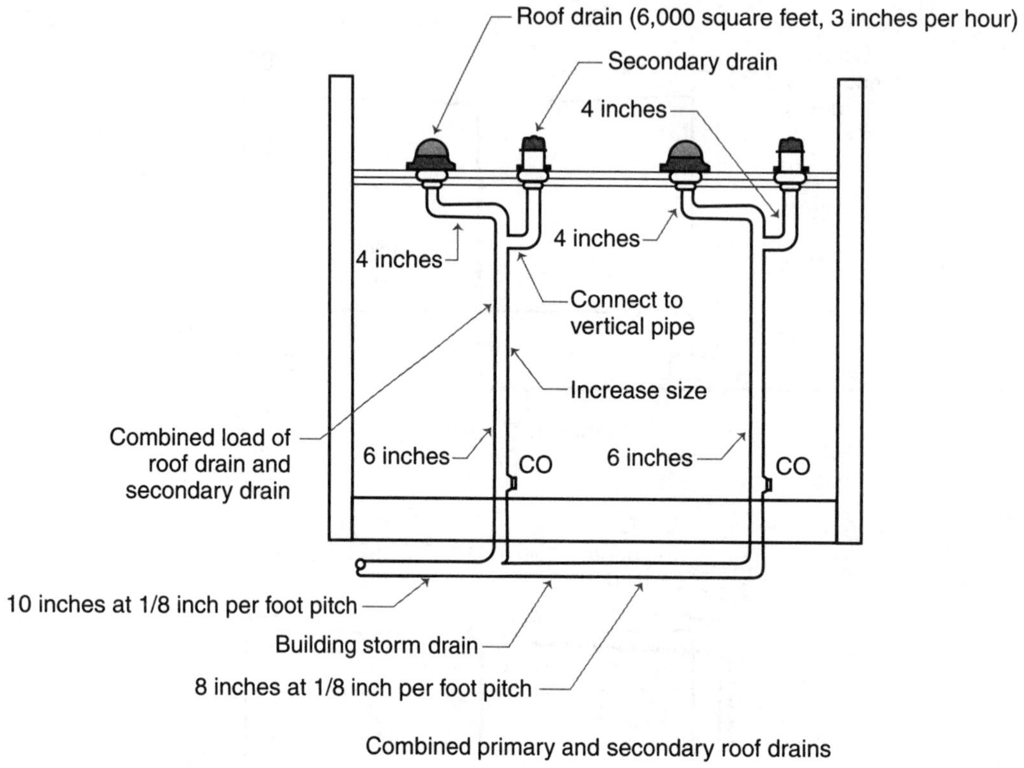

FIGURE 9.3 100-year, 1-hour rainfall (inches), eastern United States.

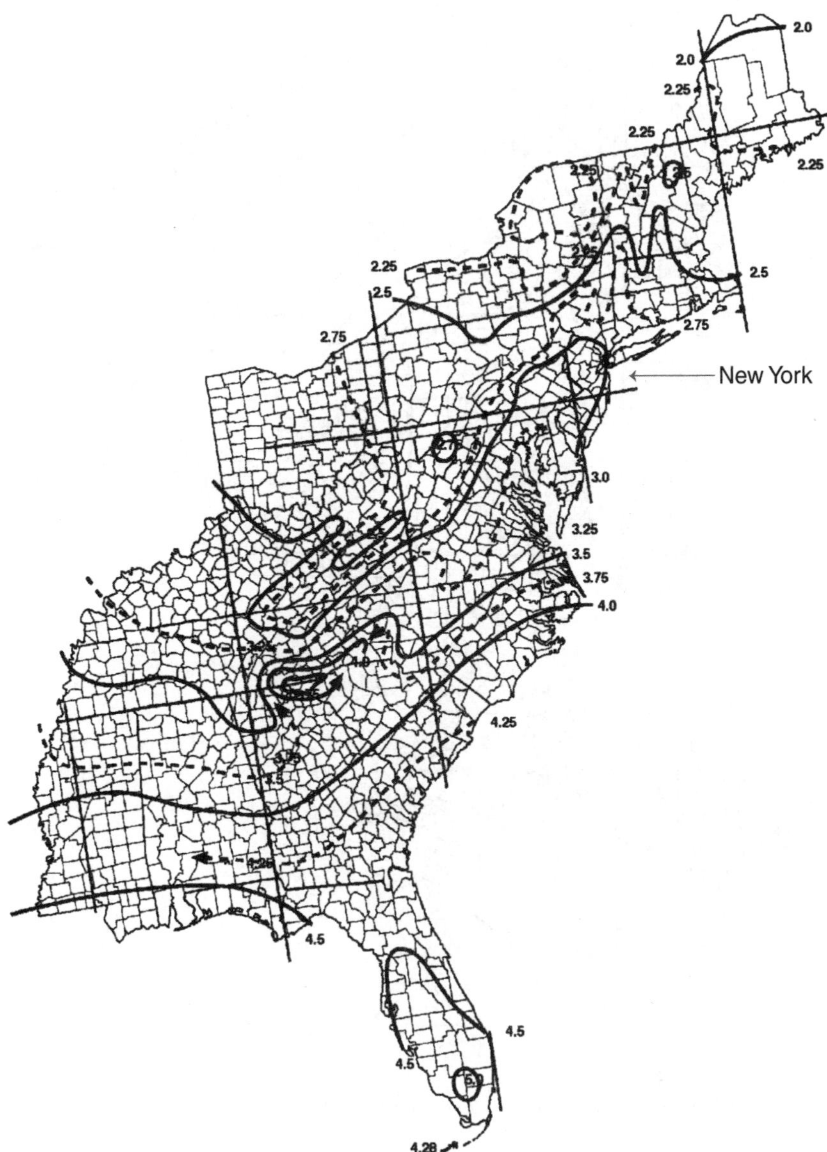

For SI: 1 inch = 25.4 mm.

(*National Weather Service, National Oceanic and Atmospheric Administration, Washington, DC*)

FIGURE 9.4 100-year, 1-hour rainfall (inches), central United States.

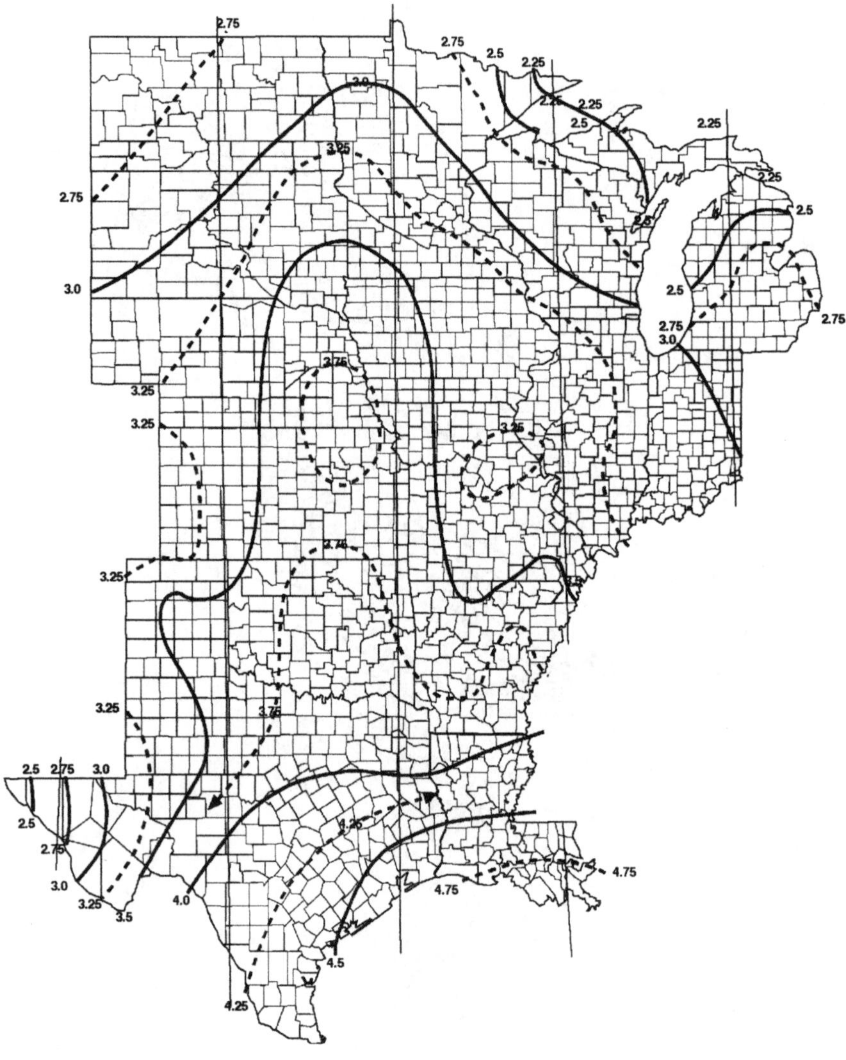

For SI: 1 inch = 25.4 mm.
(*National Weather Service, National Oceanic and Atmospheric Administration, Washington, DC*)

FIGURE 9.5 100-year, 1-hour rainfall (inches), western United States.

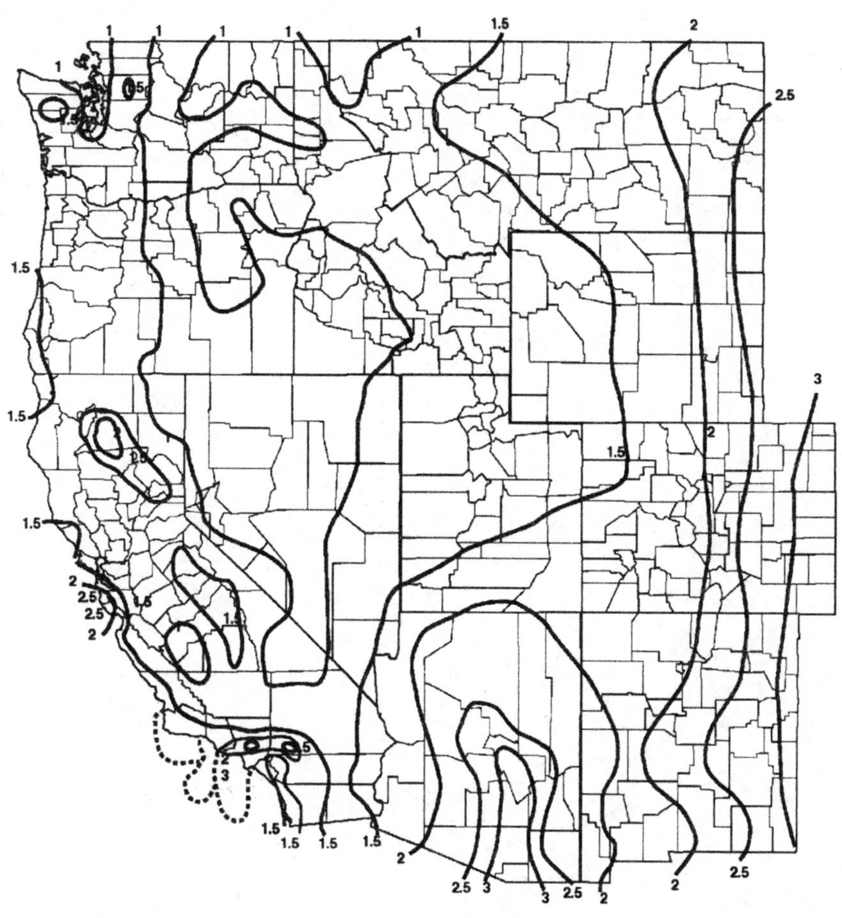

For SI: 1 inch = 25.4 mm.

(*National Weather Service, National Oceanic and Atmospheric Administration, Washington, DC*)

—NOTES—

Chapter 10

TRAPS, CLEANOUTS, AND SPECIAL WASTE

The purpose of a trap is to keep sewer gases from being emitted out of any drainage system. This is accomplished by a nonmechanical liquid seal in the pipe itself consisting of a 180-degree section of the discharge pipe. Sewer gas frequently contains hydrogen sulfide and other odorous and flammable gases that could cause an explosion if there is an ignition source. A trap is intended to be simple, must have a smooth interior, and is constructed of materials that are resistant to erosion and corrosion. It shall have a smooth interior pattern that offers minimal resistance to the passage of water and allows water to pass unobstructed through it. It will not be prone to collect any crud or gunk. Traps must be set to an interior retention depth of water seal that will reduce the possibility of self-siphoning and the release of any gas that is objectionable or a hazard to building occupants. The various types of traps included in the code are illustrated in Figures 10.1 through 10.5. A complete installation of a floor cleanout is shown in Figure 10.6.

The purpose of a cleanout is to provide access to the inside of a pipe for maintenance and removal of stoppages where they occur. Cleanouts are required to be installed at the base of vertical stacks where the pipe changes direction from vertical to horizontal, at changes of direction in horizontal pipe greater than 45 degrees, and along horizontal runs of 100 feet in length. A typical complete installation of a "P" trap is illustrated in Figure 10.7.

Special wastes contain hazardous material. They are generated by fixtures that are not traditionally plumbing fixtures but rather from equipment and other devices that are considered specialized and from commercial and industrial sources. These hazardous waste components must be separated and removed from ordinary waste before special waste is discharged into the sanitary drainage system. This is covered in Chapter 11.

MULTIPLE-CHOICE EXAM

1. For building sewers 8 inches (DN 200) in diameter and larger, a method of cleanout is *not* required at which of the following?

   a. At changes in direction

   b. At intervals of not more than 400 feet (120 m)

   c. At intervals of not more than 100 feet (30 m)

   d. 100 feet (30 m) from the junction of the building drain and the building sewer

2. Corrosive or other harmful chemicals spilling into the plumbing system shall *not* be

   a. thoroughly diluted.
   b. neutralized.
   c. treated.
   d. treated by methods chosen by the owner.

3. Which one of the following wastes may have a harmful effect on plumbing or waste-disposal systems?

   a. Special waste
   b. Water waste
   c. Sewer waste
   d. Indirect waste

4. Water exceeding what temperature is *not* permitted to drain directly into a drainage system?

a. 110°F (43°C) b. 120°F (49°C)

c. 140°F (60°C) d. 210°F (99°C)

5. What is the permitted vertical distance from a fixture outlet to a trap inlet?

a. 12 inches (300 mm) b. 18 inches (450 mm)

c. 24 inches (600 mm) d. 30 inches (750 mm)

6. What is the permitted horizontal distance from a fixture outlet to the trap weir?

a. 12 inches (300 mm) b. 18 inches (450 mm)

c 24 inches (600 mm) d. 30 inches 30 inches (750 mm)

7. What type of trap is approved for installation on a dental cuspidor to intercept solids?

a. Bell b. Drum

c. Crown vented d. S trap

8. For a trap serving a single or combination fixture, the developed length of the most upstream fixture to the inlet of the receptor shall *not* exceed which of the following?

a. 24 inches (600 mm) b. 30 inches (750 mm)

c. 60 inches (1,500 mm) d. 72 inches (1,800 mm)

9. What type of traps is prohibited?

a. A bell trap b. A drum trap used as a solids interceptor

c. A crown vented trap d. An S-type trap

10. For a building sewer provided with a house trap and a fresh-air inlet, the fresh-air inlet is *not* required to comply with which of the following?

a. Be the same size as the building sewer

b. Be half the size of the building sewer

c. Be carried above grade outside the building

d. Have a protected opening

11. The size of a trap seal for a P trap shall *not* be less than which of the following values?

a. 1 inch (25 mm) b. 1½ inches (38 mm)

c. 2 inches (50 mm) d. 2½ inches (63 mm)

12. Where shall slip joints connecting traps to the drainage system be located?

a. At the immediate trap inlet
b. At the immediate trap outlet
c. Within the trap seal
d. Directly at the fixture tailpiece

13. When provided in a parking structure, what is the requirement for floor drains connecting to a combined building sewer system?

a. They must be individually trapped.
b. They must be connected through a separator.
c. They must be connected directly to the drainage system.
d. They must be connected into a main trap.

14. The water discharged from an emergency shower shall

a. be classified as hazardous waste.
b. spill into a containment area.
c. be treated as hazardous waste.
d. be required to have a waste connection.

15. A water supply or discharge at a combined emergency eye and face wash station

a. can be supplied with tepid water.
b. cannot be classified as hazardous waste.
c. can be supplied with cold water.
d. shall be classified as hazardous waste.

16. The code refers to emergency showers and eyewashes that must conform to Standard ISEA Z358.1. What is the title of this standard?

a. Installation of Emergency Showers
b. Installation of Emergency Eyewashes
c. Emergency Eyewash and Shower Equipment
d. Emergency Showers and Eyewashes

17. How often shall cleanouts be provided on interior horizontal building sewers of 6 inches and smaller in diameter?

a. Every 125 feet (37.5 m)
b. Every 100 feet (30 m)
c. Every 75 feet (22.5 m)
d. Every 50 feet (15 m)

18. Cleanout plugs on metallic piping shall be made of which of the following materials?

a. Lead
b. Copper
c. Steel
d. Brass

19. How large shall clearance be for cleanouts on 6-inch-diameter and smaller piping for the purpose of rodding?

a. 12 inches (300 mm)
b. 18 inches (450 mm)
c. 24 inches (600 mm)
d. 30 inches (750 mm)

20. What is the smallest size allowed for cleanouts in piping larger than 8 inches in diameter (DN 200)?

a. 4 inches (100 mm)
b. 3 inches (75 mm)
c. 6 inches (150 mm)
d. 8 inches (200 mm)

21. What is a fitting or assembly installed in a main building drain for the purpose of improving the circulation of air between the drain and the sewer called?

a. A backwater valve
b. An interceptor
c. A separator
d. A house trap

22. A combination plumbing fixture is permitted to have a single trap provided that one compartment is deeper than the other by what dimension?

a. 4 inches (100 mm)
b. 6 inches (150 mm)
c. 8 inches (200 mm)
d. 12 inches (300 mm)

23. A combination plumbing fixture is permitted to have a single trap provided that one compartment has the waste outlet separated by no more than how many inches?

a. 12 inches (300 mm)
b. 24 inches (600 mm)
c. 30 inches (900 mm)
d. 36 inches (900 mm)

24. For building sewers 8 inches (DN 200) and larger in diameter, what is the maximum interval manholes shall be allowed?

a. 100 feet (30 m)
b. 200 feet (60 m)
c. 300 feet (90 m)
d. 400 feet (120 m)

25. Where is a domestic washing machine *not* permitted to discharge waste?

a. Through an air gap
b. Past the trap of a kitchen sink
c. Through an air break
d. Into a Y branch on a sink tailpiece

26. What is the maximum water seal permitted on a fixture trap?

a. 2 inches (50 mm)
b. $2\frac{1}{2}$ inches (62.5 mm)
c. 3 inches (75 mm)
d. 4 inches (200 mm)

FIGURE 10.1 Typical fixture trap.

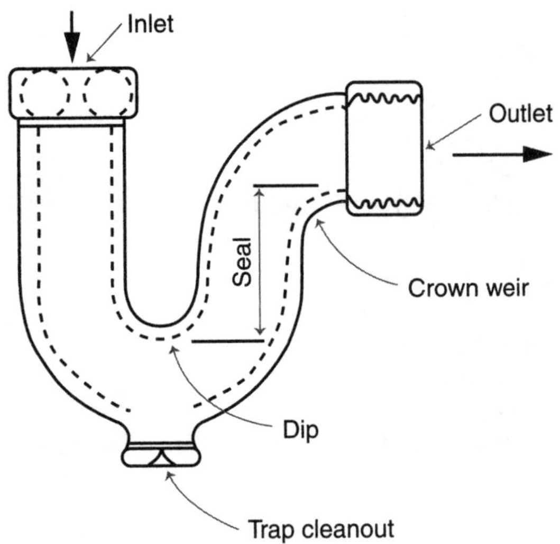

FIGURE 10.2 Crown vented trap.

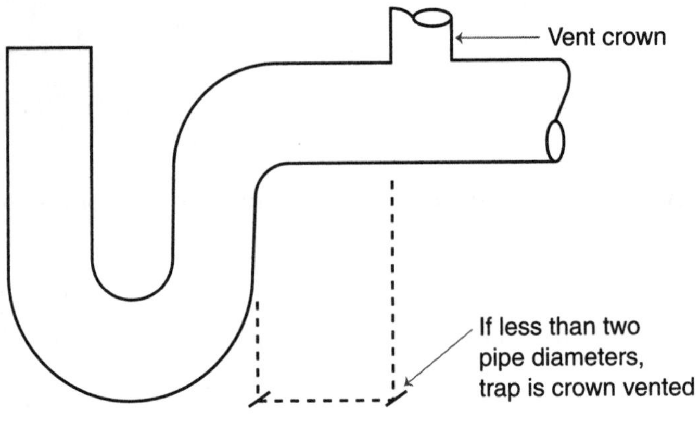

FIGURE 10.3 Drum trap.

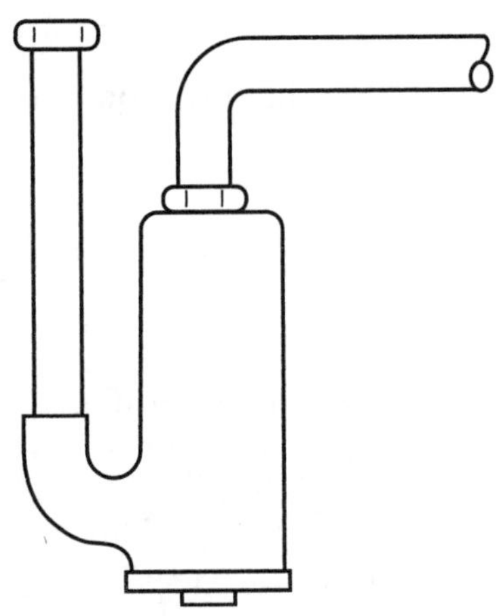

FIGURE 10.4 Full S trap.

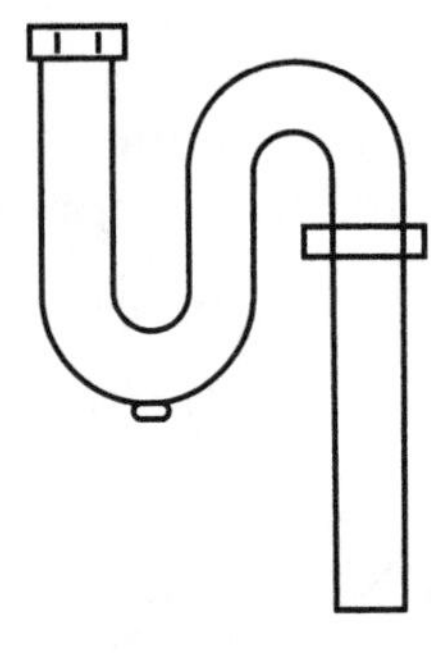

FIGURE 10.5 Typical hair and solids interceptor. (*Courtesy of Jay R. Smith Company.*)

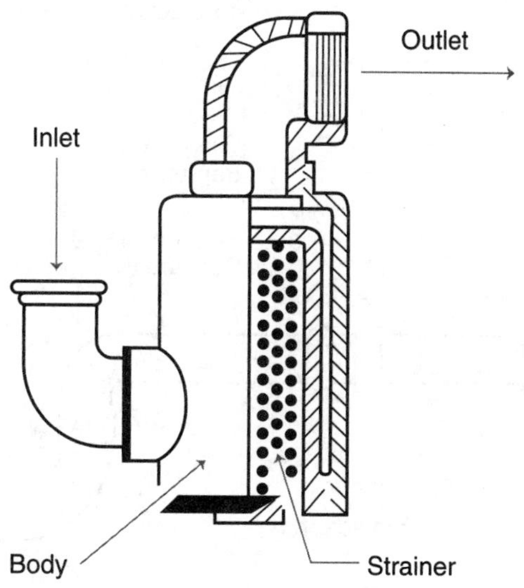

FIGURE 10.6 (a) Typical cleanouts; (b) cleanout components. (*Courtesy of Jay R. Smith Company.*)

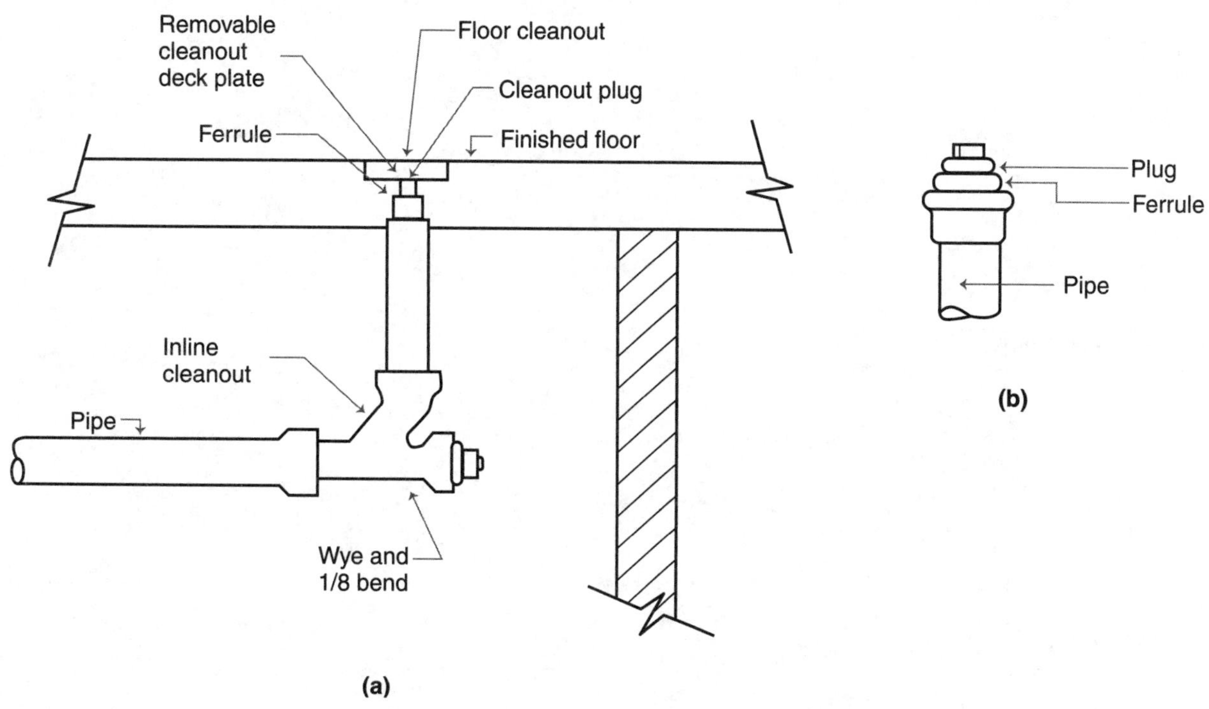

FIGURE 10.7 Installation of a P trap.

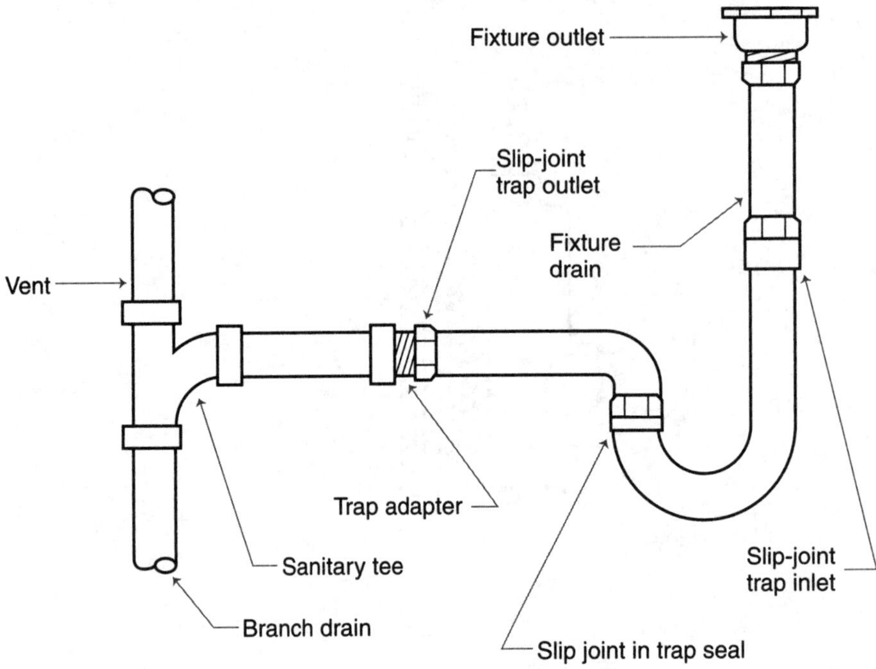

Chapter 11

INTERCEPTORS AND SEPARATORS

An *interceptor* is an inline device that prevents contaminated wastewater in a drainage stream from discharging into the sanitary system, site, or public sanitary sewers. The nature of this contamination is determined by code officials and is different for various jurisdictions. It is important that this contamination be intercepted, neutralized, separated, and disposed of or recovered from the sanitary waste discharge prior to being routed into the sanitary system.

The code states that "corrosive liquids, spent acids and other harmful chemicals that destroy or injure a drain, sewer, soil or waste pipe or create noxious or toxic fumes or interfere with the sewage treatment process shall not be discharged into the plumbing system without being thoroughly diluted, neutralized or treated by passing through an approved dilution or neutralizing device." Dilution alone is not the preferred manner of treatment. The preferred treatment is neutralization with an acid neutralizer such as that illustrated in Figure 11.1, using limestone chips as the neutralizing agent.

Another type of contamination is from fats, oils, and grease (FOG), which are a normal part of the typical waste water from a facility with a kitchen. They are removed by a grease (or FOG) interceptor. A typical FOG interceptor is illustrated in Figure 11.2. *Industrial wastewater* is a generic term used to describe nonsanitary (plumbing) effluent, such as chemicals and volatile oil that typically are found in chemical, pharmaceutical, industrial, and other manufacturing facilities other than food establishments. This type of oil interceptor is illustrated in Figure 11.3. Sand, such as found in a footing drain system, is separated from the waste stream by a sand interceptor, such as that illustrated in Figure 11.4. Solids from jewelry establishments and other particulates that may be in the drainage stream are salvaged in drum traps with added screens, such as that illustrated in Figure 10.5.

Interceptors and separators also may discharge wastewater from commercial facilities such as self-service laundries and large restaurants. The code definition of such liquids also includes storm water runoff containing anything considered harmful by a code official.

MULTIPLE-CHOICE EXAM

1. A device that is designed and installed to prevent the discharge of hazardous or undesirable substances is known as what?

   a. A backwater valve
   b. An interceptor
   c. A drum trap
   d. A screen

2. Grease interceptors must be equipped with devices to control the rate of which of the following?

   a. Air flow
   b. Solids flow
   c. Heat transference
   d. Water flow

3. Oil separators must have a depth of at least how many feet below the invert of a discharge drain?

   a. 2 feet (0.60 m)
   b. 4 feet (1.2 m)
   c. 5 feet (1.5 m)
   d. 3 feet (0.9 m)

4. The outlet opening of an oil separator must have a water seal of not less than how many inches:

   a. 24 inches (600 mm)
   b. 22 inches (550 mm)
   c. 18 inches (450 mm)
   d. 36 inches (900 mm)

5. Oil separators that are installed in areas where automobiles are serviced must have a minimum capacity of how many cubic feet for the first 100 square feet (33 m^2) of area to be drained?

   a. 4 square feet (3.3 m^2)
   b. 6 square feet (5 m^2)
   c. 8 square feet (6.6 m^2)
   d. 12 square feet (10 m^2)

6. After the first 100 feet (33 m), oil separators must be enlarged at a rate of how many cubic feet for each additional 100 square feet (33 m^3) of area to be drained?

   a. 4 cubic feet (3.3 m^3)
   b. 3 cubic feet (2.5m^3)
   c. 2 cubic feet (1.6 m^3)
   d. 1 cubic feet (.83 m^3)

7. Sand interceptors must be installed so that they are

   a. convenient.
   b. concealed.
   c. readily accessible.
   d. available.

8. Sand interceptors must have water seals with depths of not less than how many inches?

   a. 6 inches (150 mm)
   b. 10 inches (250 mm)
   c. 14 inches (350 mm)
   d. 18 inches (450 mm)

9. Interceptors used in commercial laundries must be sized to prohibit the entrance of solids with what diameter or larger from entering the drainage system?

   a. ¼ inch (6.5 mm)
   b. ½ inch (13 mm)
   c. ¾ inch (18.5 mm)
   d. 1 inch (25 mm)

10. Commercial laundries must be equipped with an interceptor containing which of the following?

    a. A wire basket capable of removal
    b. A solid cover
    c. A cover that is permanently attached
    d. An integral vent

11. The interceptors used in bottling establishments must be capable of separating which of the following materials from other solids before discharging the waste into the drainage system?

    a. Flammable materials
    b. Broken glass
    c. Oil
    d. Bottled fluids

12. Separators used in slaughterhouses are not required to prevent which of the following substances from entering the drainage system?

   a. Feathers

   b. Entrails

   c. Material that may cause stoppages

   d. FOG

13. What must be done if an interceptor or separator is subject to loss of trap seal by evaporation?

   a. It must be provided with a trap primer.

   b. It must be vented.

   c. It must be inspected regularly.

   d. It must be drained of all air.

14. How often should interceptors be cleaned?

   a. Periodically
   b. Every day
   c. Once a week
   d. Once a month

15. The installation of a grease interceptor manufactured to serve as a fixture trap is limited to which of the following?

   a. No more than two fixture drains

   b. A maximum of 60 inches (1,500 mm) from the fixture to the outlet of the interceptor

   c. A maximum of 30 inches (750 mm) horizontally from the fixture to the outlet of the interceptor

   d. A maximum of 24 inches (600 mm) horizontally from the fixture to the outlet of the interceptor

16. For a grease trap serving a single or combination fixture, how long should the vertical distance to the inlet of the receptor be?

   a. 30 inches (750 mm)
   b. 24 inches (600 mm)
   c. 18 inches (450 mm)
   d. 12 inches (300 mm)

17. A grease interceptor is permitted to be used as a single trap for a combination fixture if the vertical distance to the inlet does not exceed which of the following values?

   a. 30 inches (750 mm)
   b. 24 inches (600 mm)
   c. 18 inches (450 mm)
   d. 12 inches (300 mm)

18. When a grease interceptor is permitted to be used as a single trap for a combination fixture, the maximum horizontal developed length from the most upstream compartment to the inlet of the interceptor must not exceed which of the following values?

a. 60 inches (1,500 mm)
b. 72 inches (1,800 mm)
c. 48 inches (1,200 mm)
d. 36 inches (900 mm)

19. What type of trap is approved for installation on a dental cuspidor to intercept solids?

a. Bell
b. Drum
c. Crown vented
d. S trap

20. For which of the following is a grease interceptor *not* required?

a. Hospitals
b. Factories
c. Schools
d. Residential units

21. Based on what should automatic grease-removal devices be sized?

a. Peak usage flow
b. Flow from all connected fixtures
c. Average usage flow
d. Flow from highest-volume fixture

22. The minimum height for a vent termination on a grease interceptor flow control device shall be 6 inches (150 mm) above which of the following?

a. Building roof
b. Interceptor flood rim
c. Connected-fixture flood rim
d. Interceptor drain connection

23. An acid-resistant trap installed in concrete underground shall be embedded to what distance beyond the sides and bottom of the trap?

a. 2 inches (50 mm)
b. 6 inches (150 mm)
c. 3 inches (75 mm)
d. 8 inches (200 mm)

24. Oil separators shall have a minimum depth of what distance below the invert of the discharging drain?

a. 24 inches (600 mm)
b. 30 inches (750 mm)
c. 36 inches (900 mm)
d. 48 inches (1,200 mm)

FIGURE 11.1 Small acid neutralizing sump. (*Source: Schier Products.*)

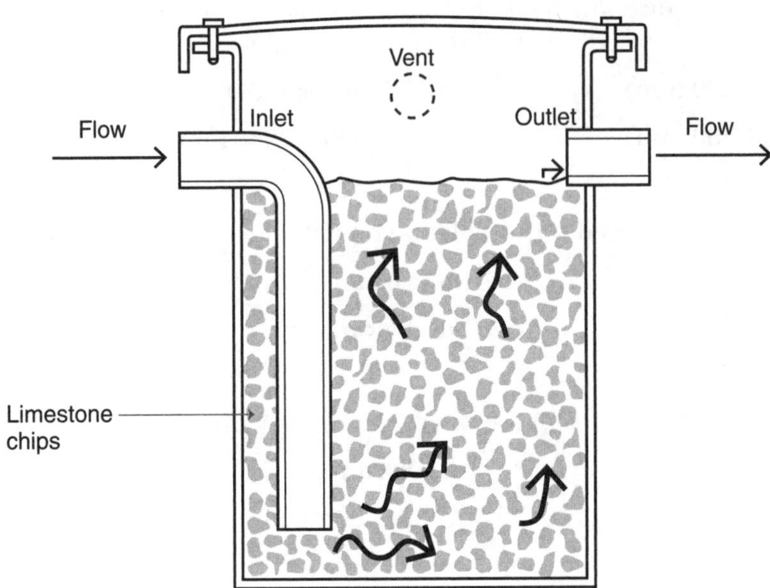

FIGURE 11.2 Typical FOG interceptor.

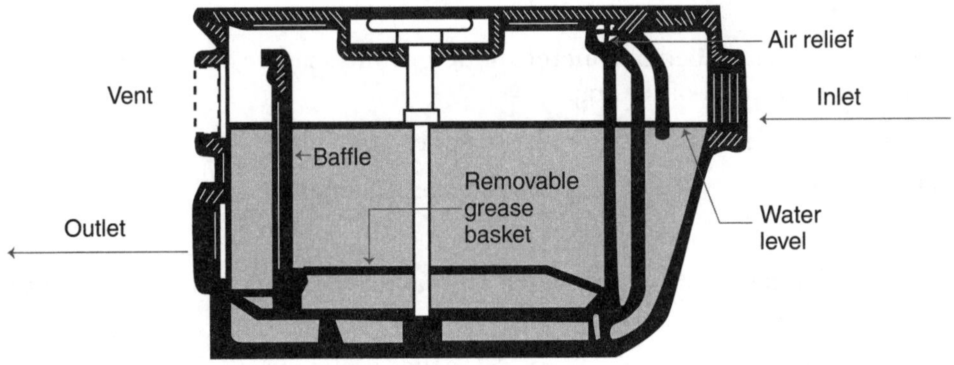

FIGURE 11.3 Typical oil interceptor. (*Courtesy of Rockford Company.*)

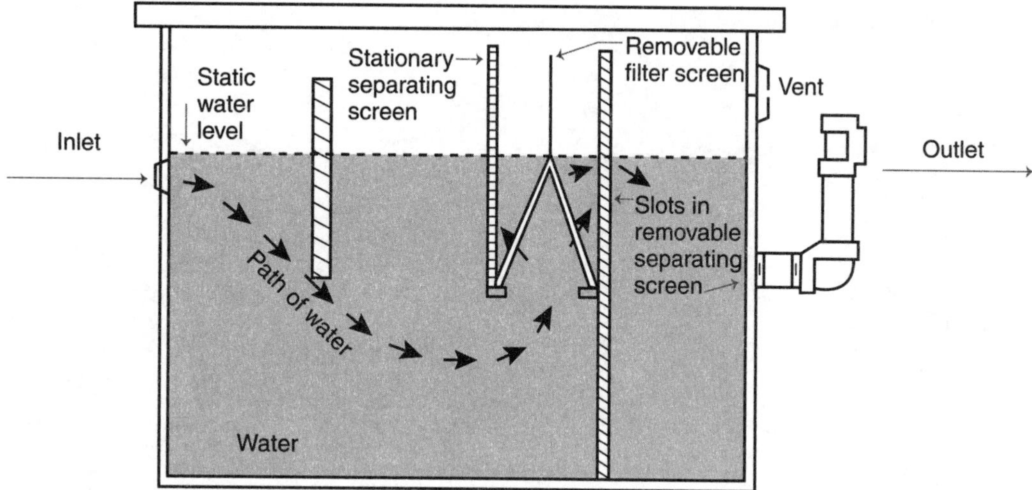

FIGURE 11.4 Sand interceptor.

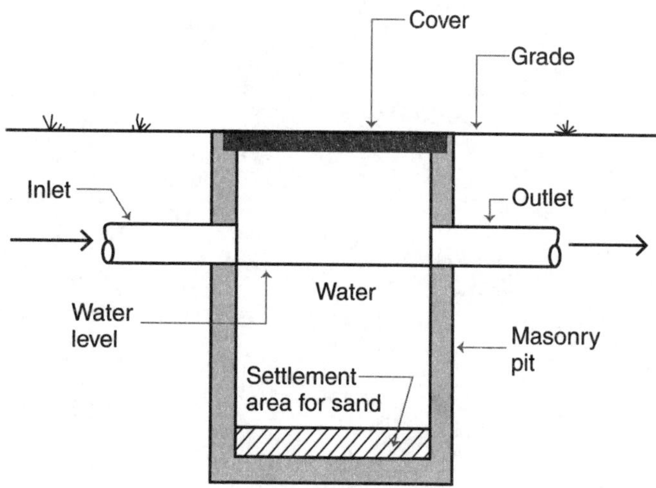

—NOTES—

Chapter 12

MEDICAL AND NONMEDICAL GASES

Chapter 12 of the *IPC* is titled, "Special Piping and Storage Systems," and is very short because it only refers to another code, National Fire Protection Association (NFPA)–99C, 2005 edition, *Gas and Vacuum Systems for Healthcare Facilities.* It is intended to apply only to the plumbers who install nonflammable medical gas systems in any facility. The category for installing medical gases is a specialized one that requires passing a test on the *Professional Qualification Standard for Medical Gas Installers.* This is a separate test required for all installers of medical gas systems and is not part of the plumbing test.

In the 2012 edition of the *IPC*, reference is made to NFPA-99C 2005, which is a reproduction of only the piping portion of the entire NFPA-99 Standard. As of 2011, this section was discontinued and was no longer available. If reference to this portion of the code is desired, the only recourse is to obtain the entire 2012 edition of NFPA-99, *Health Care Facilities Code,* which is not referenced.

There should be no questions pertaining to this section on any test addressing the *IPC* because it refers to information not included in the code. However, there is reference to one aspect: In the 2012 edition of the *UPC*, there is an entire section concerning health-care facilities and much reference to NFPA-99, *Health Care Facility Code,* which is not an included part of the test for plumbing. For the most part, all portions of the UPC refer to NFPA-99.

MULTIPLE-CHOICE EXAM

1. Using the *IPC*, this chapter does not apply for
   a. medical gas systems installed in hospitals.
   b. medical gas systems installed in nursing homes.
   c. medical gas systems installed in dentist offices.
   d. portable systems or cylinder storage.

Chapter 13

ALTERNATIVE ENGINEERED AND SPECIAL SYSTEMS

Alternative engineered systems have been included in codes to allow drainage and vent systems that use piping, systems, and equipment employing other than traditional methods. Every jurisdiction is allowed to adapt and revise any code that is adopted. If some of them are progressive enough to allow any nontraditional systems, officials are willing to accept a professional engineer's design if it will work, at least on paper. Some alternative systems are included in every code and others are not. If the local area where the test is to be taken does not use the *IPC*, portions of this chapter may be skipped because these items will not appear on that test. If a system is designed and approved by a professional engineer, sufficient information must be included in the contract documents to allow a complete review by the code officials. Recently developed computerized systems use methods and theory that have been developed for the computer software.

There are several sanitary and vent systems to be discussed here, namely the single-stack systems, recently added to the 2012 *IPC* and referred to in the code as "single-stack vent systems," (commonly known as *Philadelphia systems*); the proprietary Sovent System, classified as a vent system and not referred to in any code; and engineered vent systems. Not all of these systems are included in all codes.

There is no adequate explanation for why these systems have not been universally accepted by the various code bodies. All of them have been in successful operation for many years, are installed throughout the country and the world, and have a very successful history. All have a potential for substantial cost savings compared with a conventional two-pipe installation.

SINGLE-STACK SYSTEM

The single-stack system in a prior edition to the 2012 *IPC* is considered to be an alternative engineered design. It consists of a combination drainage and vent system that provides only one pipe for the flow of liquid, solids, and air without the loss of trap seals. This system is based on the *Philadelphia Plumbing Code*, which has been used for more than 100 years. If the test that is being given is based on an earlier edition than the 2012 *IPC*, this chapter is considered to address an alternative engineered design, which requires that the design and sizing be done by a registered professional engineer and thus will not be included as a code requirement. However, this system has been added to the 2012 edition of the *IPC* as a code requirement.

In general, the stack and vertical drops from fixtures are oversized, and the lengths of trap arms are limited. Fixture drains that do not meet code requirements shall be vented conventionally. The following questions are based on the single-stack system included in the vent section of the 2012 *IPC*. Thus the questions are based only on alternative plumbing systems.

MULTIPLE-CHOICE EXAM

1. Based on a single-stack design, how should the drainage stack be sized?

a. Uniformly throughout its length

b. Increased in size based on connected fixture units at each branch interval

c. Based on developed length

d. Based only on water closets connected

2. Based on a single-stack design, water closet connections to a stack shall *not* be greater than what developed length?

a. 2 feet (0.6 m) b. 3 feet (0.9 m)

c. 4 feet (1.2 m) d. 5 feet (1.5 m)

3. Based on a single-stack design, fixtures other than water closet connections to a stack shall *not* be greater than what developed length?

a. 6 feet (1.8 m) b. 8 feet (2.4 m)

c. 10 feet (3 m) d. 12 feet (3. 6 m)

4. Based on a single-stack design, water closet connections to a stack made through a sanitary tee shall *not* be greater than what developed length?

a. 6 feet (1.8 m) b. 8 feet (2.4 m)

c. 10 feet (3 m) d. 12 feet (3.6 m)

5. Based on a single-stack design, fixtures other than water closets shall not be located more than what distance from a stack?

a. 8 feet (2.4 m) b. 10 feet (3 m)

c. 12 feet (3.6 m) d. 15 feet (4.5 m)

6. Based on a single-stack design, the maximum vertical drop into a horizontal branch shall be which of the following?

a. 2 feet (0.6 m) b. 4 feet (1.2 m)

c. 6 feet (0.9 m) d. 8 feet (2.4 m)

7. Based on a single-stack design, fixture drains that have a vertical drop in excess of what distance shall be individually vented?

a. 4 feet (1.2 m) b. 6 feet (0.9 m)

c. 8 feet (2.4 m) d. 10 feet (3 m)

8. Based on a single-stack design, stacks higher than two branch intervals shall have no connections where?

a. On the lower three floors
b. No connections are allowed.
c. On the lower one floor
d. On the lower two floors

9. Based on a single-stack design and using Table 13.1, how many drainage fixture units (DFUs) are allowed for a 5-inch (DN 125) single stack that is less than 75 feet (22.5 m) in height?

a. 24
b. 225
c. 480
d. 1,025

10. How many DFUs are allowed for a single-stack building sewer or building drain for a system 6 inches (150 mm) in size and sloping at ¼ inch per foot (2 cm/m) using Table 13.2?

a. 700
b. 840
c. 1,000
d. 480

TABLE 13.1 Single-Stack Sizes

STACK SIZE (INCHES)	MAXIMUM CONNECTED DRAINAGE FIXTURE UNITS		
	STACKS LESS THAN 75 FEET IN HEIGHT	STACKS 75 FEET TO LESS THAN 160 FEET IN HEIGHT	STACKS 160 FEET AND GREATER IN HEIGHT
3	24	NP	NP
4	225	24	NP
5	480	225	24
6	1,015	480	225
8	2,320	1,015	480
10	4,500	2,320	1,015
12	8,100	4,500	2,320
15	13,600	8,100	4,500

For SI: 1 inch = 25.4 mm; 1 foot = 304.8 mm.

TABLE 13.2 Building Drains and Sewer Sizes

DIAMETER OF PIPE (INCHES)	MAXIMUM NUMBER OF DRAINAGE FIXTURE UNITS CONNECTED TO ANY PORTION OF THE BUILDING DRAIN OR THE BUILDING SEWER, INCLUDING BRANCHES OF THE BUILDING DRAIN			
	SLOPE PER FOOT			
	1/16 INCH	1/8 INCH	1/4 INCH	1/2 INCH
1¼	—	—	1	1
1½	—	—	3	3
2	—	—	21	26
2½	—	—	24	31
3	—	36	42	50
4	—	180	216	250
5	—	390	480	575
6	—	700	840	1,000
8	1,400	1,600	1,920	2,300
10	2,500	2,900	3,500	4,200
12	3,900	4,600	5,600	6,700
15	7,000	8,300	10,000	12,000

For SI: 1 inch = 25.4 mm.

—NOTES—

Chapter 14

GRAY-WATER RECYCLING SYSTEMS

Gray water is defined as untreated wastewater from both commercial and domestic activities. This is an increasingly used water resource whose use is now mandated in many locations.

Gray water may receive wastewater from specific individual residential systems, such as bathtubs, showers, laundry rooms, and lavatory sinks. Commercial systems are those which may receive wastewater from clothes washers. Laundry discharges from hotels and health-care facilities are also acceptable.

This chapter includes provisions that govern the design, construction, materials, and installation of gray-water systems that appear for the first time in the 2012 edition of the *IPC*. If the local area where the test is to be taken does not use the *IPC*, this chapter may be skipped because these items will not appear on the test.

MULTIPLE-CHOICE EXAM

1. Gray-water recycling systems are not permitted to receive waste from which of the following?

a. Showers
b. Baths
c. Clothes washers
d. Kitchen sinks

2. Gray water is not permitted to be used for which of the following?

a. Flushing water closets
b. Flushing urinals
c. Surface irrigation of plants
d. Subsurface irrigation of plants

3. A typical gray-water recycling system consists of which of the following?

a. Indirect discharge into a sanitary system
b. Direct discharge into a sanitary system
c. A gray-water reservoir
d. A filter

4. The reservoir shall be sized to retain how many hours of storage?

a. 24 hours
b. 18 hours
c. 12 hours
d. 8 hours

5. What color shall the gray water piping itself use?

a. Blue
b. Red
c. Yellow
d. Green

6. Disinfection is *not* permitted using which of the following chemicals?

a. Chlorine
b. Iodine
c. Ozone
d. Hydrogen peroxide

7. What is the total estimated flow demand per day from residential occupants?

a. 25 gallons per day (94.6 L/d)
b. 15 gallons per day (56.7 L/d)
c. 40 gallons per day (151 L/d)
d. Flow based only on adults

8. Disinfection is required for what use of gray water?

a. Subsurface irrigation
b. Flushing of urinals
c. Flushing water closets
d. Use in laundry sinks

9. How big shall the distribution piping diameter be for gray water?

a. Not less than 3 inches (75 mm)
b. Not less than 2½ inches (63 mm)
c. Not less than 2 inches (50 mm)
d. Not less than 4 inches (100 mm)

10. The top of the distribution piping for gray water on a site shall *not* be less than which of the following dimensions from the original ground level?

a. 10 inches (250 mm)
b. 8 inches (200 mm)
c. 12 inches (300 mm)
d. 18 inches (450 mm)

11. The seepage trench excavation shall *not* be less than what width?

a. 18 inches (450 mm)
c. 24 inches (600 mm)
c. 6 inches (150 mm)
d. 12 inches (300 mm)

12. Individual seepage trenches excavation shall *not* be more than how many feet apart?

a. 150 feet (45 m)
b. 125 feet (37.5 m)
c. 100 feet (30 m)
d. 75 feet (22.5 m)

13. The length of distribution piping in seepage beds shall *not* be greater than which of the following?

a. 5 feet (1.5 m)
b. 6 feet (1.8 m)
c. 8 feet (2.4 m)
d. 10 feet (3 m)

14. What should be the depth of aggregate below the distribution piping elevation in a trench?

a. 4 inches (100 mm)
b. 6 inches (150 mm)
c. 8 inches (200 mm)
c. 12 inches (300 m)

15. What shall be the depth of aggregate over the top of the distribution piping in a trench?

a. 8 inches (200 mm)
b. 6 inches (150 mm)
c. 4 inches (100 mm)
d. 2 inches (50 mm)

Chapter 15
NATURAL GAS

The *IPC* has a separate volume on natural gas that will not be part of the plumbing code test. The inclusion of natural gas piping on the test depends on the locality where the test is given. If it is included, applicants will have to purchase the fuel gas code for study. If you are in an area where the fuel gas has not been adopted, this chapter may be skipped.

This chapter is written for tests based on the *UPC*. For areas that have adopted the *UPC*, the test references National Fire Protection Association (NFPA–54), the *National Fuel Gas Code*, for a greater majority of its requirements. It is not necessary for applicants to have an intimate knowledge of the NFPA code, only the information appearing within the *UPC* itself.

MULTIPLE-CHOICE EXAM

1. For tubing installed inside hollow walls without protection, a striker steel barrier shall be installed between the tubing, and the wall shall extend not less than how many inches beyond the penetration?

a. 2 inches (50 mm) b. 3 inches (75 mm)
c. 4 inches (100 mm) d. 6 inches (150 mm)

2. What is the maximum design operating pressure for normal residential natural gas piping located inside a building?

a. 5 psi (35 kPa) b. 4 psi (28 kPa)
c. 3 psi (21 kPa) d. 10 psi (70 kPa)

3. Gas piping shall *not* be sized in accordance with which of the following?

a. Pipe sizing tables in this chapter.
b. Pipe sizing tables in NFPA-54.
c. Pipe sizing in manufacturing tables
d. Pipe sizing tables supplied by a contractor

4. Metallic pipe fittings larger than what size shall *not* be used indoors unless approved?

a. 2 inches (50 mm) b. 3 inches (125 mm)
c. 4 inches (100 mm) d. 6 inches (150 mm)

5. Overpressure protection shall limit the downstream pressure from the line pressure regulator to what value?

a. 1 psi (7 kPa) b. 2 psi (14 kPa
c. 3 psi (21 kPa) d. 4 psi (28 kPa)

6. A regulator vent shall terminate not less than how many feet from a source of ignition?

   a. 3 feet (1 m) b. 4 feet (1.2 m)

   c. 5 feet (1.5 m) d. 6 feet (2 m)

7. Under normal conditions, what depth underground shall a gas pipe be installed?

   a. 12 inches (300 mm) b. 18 inches (450 mm)

   c. 24 inches (600 mm) d. 30 inches (750 mm)

8. An unthreaded portion of a gas piping outlet shall extend how many inches above the surface of a floor?

   a. 1 inch (25 mm) b. 1½ inches (28 mm)

   c. 2 inches (50 mm) d. 3 inches (75 mm)

9. An unthreaded portion of a gas piping outlet shall extend how many inches through the surface of a finished ceiling?

   a 1 inch (25 mm) b. 1½ inches (28 mm)

   c. 2 inches (50 mm) d. 3 inches (75 mm)

10. An indoor nonmetallic gas hose connector shall *not* exceed what length?

    a. 2 feet (0.75 m) b. 3 feet (1 m)

    c. 4 feet (1.2 m) d. 6 feet (2 m)

11. An outdoor nonmetallic gas hose connector shall *not* exceed what length?

    a. 5 feet (1.75 m) b. 3 feet (1 m)

    c 15 feet (4.5 m) d. 20 feet (6.75 m)

12. A pressure test for a fuel gas system is not permitted to use what as a test gas?

    a. Carbon dioxide b. Liquefied petroleum gas

    c. Nitrogen d. Air

13. An indoor nonmetallic appliance shutoff valve shall *not* exceed how many feet from the appliance it serves?

    a. 2 feet (1 m) b. 3 feet (1 m)

    c. 4 feet (1.2 m) d. 6 feet (2 m)

14. The test pressure for natural gas piping shall be no less than which of the following?

    a. 2 psi (14 kPa) b. 5 psi (35 kPa)

    c. 10 psi (69 kPa) d. 15 psi (105 kPa)

15. How long shall the test pressure be maintained for natural gas piping at a pressure of 14 inches of water?

a. No less than 20 minutes
b. No less than 15 minutes
c. No less than 10 minutes
d. No less than 5 minutes

16. When purging a gas-line system with a pressure of more than 2 psi (17 kPa) when the purge pipe is extended to the outside of a building, which of the following points of discharge from the open end is not necessary to be complied with?

a. No less than 10 feet from the building
b. Monitored by a gas indicator
c. Controlled with a shutoff valve
d. Attended by any person

Example of Sizing: The following example is given only for illustration purposes.

Find the required pipe size for the piping shown in Figure 15.1.

Given: Polyethylene pipe, allowable friction loss of 0.3 inch of water, natural gas with a specific gravity 0.60. The allowable pressure loss of 0.3 inch of water is an often-selected figure for low-pressure natural gas.

Step 1: Find the gas demand of appliances from Table 15.1.

Step 2: Calculate the total run of gas piping, and select the allowable friction loss. The 0.3 inch of water loss in friction is a commonly used figure. This is for use in selecting the correct sizing chart for the pipe material that will be used.

Step 3: Find the appropriate gas sizing chart based on the subject gas (whether it is natural gas or liquefied petroleum gas), material used for the piping system, and the desired pressure loss for the piping system. Use Table 15.2.

Solution

Step 1: Water heater = 35,000 Btu
Refrigerator = 3,000 Btu
Range = 65,000 Btu

Step 2: Add all straight runs of main sections of pipe to get 30 feet. It is not necessary to use any of the branch lengths of pipe. The pressure drop commonly used is 0.3 inch of water.

Step 3: Use Table 15.2, which has the pipe material, specific gravity, and pressure drop used to calculate the pipe sizes.

Using Table 15.2 and the length of pipe column, select line 30 (feet). This is the only line to be used. All the capacity figures will be read on that line only.

For the water heater at 35,000 cfh = ½-inch pipe.

For refrigerator and heater = 38,000 cfh, read still ½-inch pipe.

For range, refrigerator plus heater = 103,000 cfh, read ¾-inch pipe.

FIGURE 15.1 Sample problem. (*Courtesy of UPC.*)

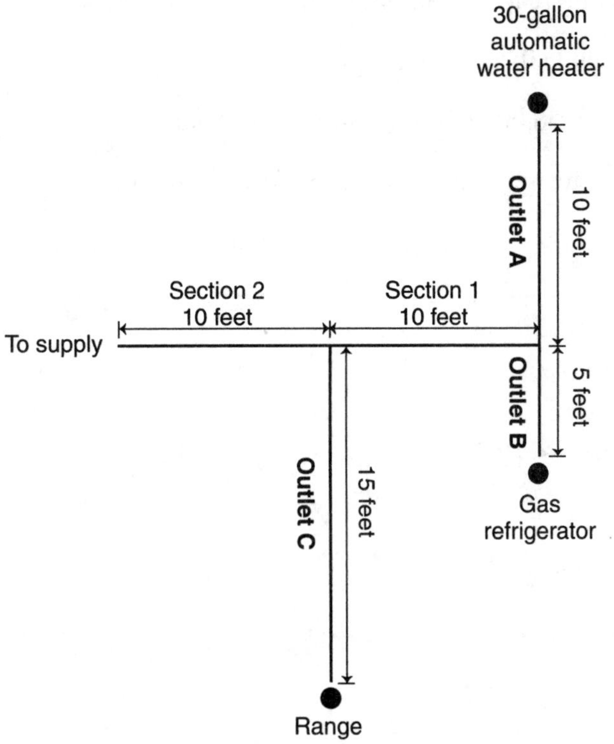

TABLE 15.1 Approximate Gas Input for Typical Appliances (*Courtesy of UPC.*)

APPLIANCE	INPUT (BTU/APPROX.)
Space Heating Units	
Warm air furnace	
Single family	100,000
Multifamily, per unit	60,000
Hydronic boiler	
Single family	120,000
Multifamily, per unit	75,000
Space and Water Heating Units	
Hydronic boiler	
Single family	120,000
Multifamily, per unit	75,000
Water Heating Appliances	
Water heater, automatic storage 30 to 40 gallon tank	35,000
Water heater, automatic storage 50 gallon tank	50,000
Water heater, automatic instantaneous	
Capacity at 2 gallons per minute	142,800
Capacity at 4 gallons per minute	285,000
Capacity at 6 gallons per minute	428,400
Water heater, domestic, circulating or side-arm	35,000
Cooking Appliances	
Range, freestanding, domestic	65,000
Built-in oven or broiler unit, domestic	25,000
Built-in top unit, domestic	40,000
Other Appliances	
Refrigerator	3,000
Clothes dryer, Type 1 (domestic)	35,000
Gas fireplace direct vent	40,000
Gas log	80,000
Barbecue	40,000
Gaslight	2,500

For SI units: 1000 British thermal units per hour = I.C.F.H.

TABLE 15.2 Polyethylene Plastic Pipe (NFPA 54-12) (*Courtesy of UPC.*)

					GAS: NATURAL			
					INLET PRESSURE: 0.3 IN. W.C.			
					SPECIFIC GRAVITY: 0.60			
	PIPE SIZE (INCH)							
NOMINAL OD:	½	¾	1	1¼	1½	2	3	4
DESIGNATION:	SDR 9.3	SDR 11	SDR 11	SDR 10	SDR 11	SDR 11	SDR 11	SDR 11
ACTUAL ID:	0.660	0.860	1.077	1.328	1.554	1.943	2.864	3.682
LENGTH (FEET)	CAPACITY IN CUBIC FEET OF GAS PER HOUR							
10	153	305	551	955	1440	2590	7170	13,900
20	105	210	379	656	991	1780	4920	9520
30	84	169	304	527	796	1430	3950	7640
40	72	144	260	451	681	1220	3380	6540
50	64	128	231	400	604	1080	3000	5800
60	58	116	209	362	547	983	2720	5250
70	53	107	192	333	503	904	2500	4830
80	50	99	179	310	468	841	2330	4500
90	46	93	168	291	439	789	2180	4220
100	44	88	159	275	415	745	2060	3990
125	39	78	141	243	368	661	1830	3530
150	35	71	127	221	333	598	1660	3200
175	32	65	117	203	306	551	1520	2940
200	30	60	109	189	285	512	1420	2740
250	27	54	97	167	253	454	1260	2430
300	24	48	88	152	229	411	1140	2200
400	21	42	75	130	196	352	974	1880
450	19	39	70	122	184	330	914	1770
500	18	37	66	115	174	312	863	1670

For SI units: 1 inch = 25 mm, 1 foot = 304.8 mm, 1 cubic foot per hour = 0.0283 m^3/h, 1 pound-force per square inch = 6.8947 kPa, 1 inch water column = 0.249 kPa

**Table entries are rounded to 3 significant digits.*

Chapter 16

GUIDE TO TAKING THE TEST

This chapter was written to assist applicants in taking the plumbing licensing exam. As you will see, there is a definite method for increasing the chances of selecting the correct answers. Both an apprentice and a master plumber are knowledgeable and experienced people with advanced installation skills that will allow them to work on any size or category of system. A master plumber's job requirements are more extensive than that of a journeyman, allowing the master to supervise other plumbers in addition to evaluating work orders, coordinating plumbing work with the other trades, and delegating work order to other employees. A master plumber has the ability to teach and employ plumbers at the journeyman's level.

The purpose of this study guide is not to teach plumbing itself. It is understood that anyone wanting to take the licensing test knows what a wrench is and how it is used. This guide will provide you with information on how to read and interpret the questions on code facts and how to answer the questions given on the test. The purpose of the test is to determine whether an individual has the necessary understanding and knowledge of various plumbing codes in different localities and the regulations necessary to work in any state. The knowledge gained by passing the test will allow individuals to train plumbers under their supervision.

The first step is to find and purchase a copy of the actual code that is used in the jurisdiction where you want to take the test. It is very important to determine the latest code year for which the test will be given. This information is given when you apply for the test. This code must be read and thoroughly understood. It is mandatory that the code be read completely from front to back again and again and again. There is no substitute for reading the code as much as possible, as boring as it may seem. Next, buy a self-help book (such as this one) that will ask technical questions allowing you to further learn the code provisions.

Along with this book, taking a class given by a union or college on the code also will help you to learn and straighten out any confusion you may have, and believe me, there will be much confusion. The cost of admission to the class will go far in helping you to understand the code. Studying with another person and asking and answering code questions back and forth will test each person's knowledge.

There is no standard test for advancement in the plumbing profession. Every state, county, and city may have a different test. Anyone thinking of taking one of these tests is assumed to already know what the plumbing profession does. It is my purpose to include enough information here that will enable readers to pass the test that you are interested in taking. Although licensing requirements and format vary across the country, most areas require some sort of a license. The amount of work experience that any individual must have to qualify differs across the states, but most states require at least four and sometimes five years of experience as a trainee or apprentice.

The first step to becoming a licensed plumber is to get certified experience in the plumbing field. This can be accomplished by working as a trainee or apprentice, which will allow you to gain work experience for at least the minimum time required by the authorities. Apprentices and trainees do not need any certification and can be hired off the street. There is usually an age requirement of 17 or 18 years. Some jurisdictions require a high school degree or GED certificate to start work. There also may be a requirement by some states for applicants to attend a trade school or take other academic courses for a certain period of time. Some states do not require an individual to be certified or licensed to work as a plumber at the journeyman level. However, it is always important to document your work experience.

Two different tests are the combined subject of this book: the journeyman plumber's licensing and the master plumber's license. The journeyman plumber's licensing exam is a state- or area-specific examination that will be used to determine whether an the individual has the necessary skills and knowledge to work as a licensed plumber and is allowed to repair, install, and maintain any plumbing system. The standard amount of experience varies, but is usually between four or five years as a trainee or apprentice. The individual must work under the supervision of a master plumber for that time. Individuals who are not required to have a journeyman's license but have met all the other requirements set by the state to work at the journeyman's level still can do the same work. Journeymen usually are not permitted to manage plumbing projects that employ a large number of other plumbers, though.

The master plumber's licensing exam is a state- or area-specific examination that is similar to the journeyman's exam. It is usually an expanded journeyman's test. The normal amount of required work experience is at least four to five years as a journeyman plumber. A master plumber's job requirements are more extensive than those of a journeyman. It is the master plumber's responsibility to evaluate work orders, coordinate and delegate work orders to other employees, train and manage employees and meet with potential clients to review plumbing problems. Often a knowledge of federal plumbing codes is necessary, as well as the financial basics of owning a business.

As a final word, it is imperative that you check with laws, rules, and regulations of the particular jurisdiction where you wish to take the test. This book is divided into various categories that will simplify complete study of the code. In order to reinforce the information contained in this book, the following list provides the basics of plumbing:

1. Plumbing involves the practice, materials, and fixtures used for the installation and maintenance of plumbing systems within or adjacent to any structure, including all public and private water supply systems (both potable and non potable); sanitary, storm water, and other liquid drainage systems; and venting, all conforming to the provisions of the local code and regulations.

2. A plumber is the qualified and licensed entity engaged in the installation, maintaining and repair of plumbing work, which includes a knowledge of the ordinances affecting the work he installs or maintains.

3. A plumbing system includes all water-distribution piping and appurtenances within a building and water-treatment equipment and all building drains, vents, fixtures, and waste piping.

4. The purpose of venting is to equalize the air pressure within a waste piping system including traps. A trap is a device intended to prevent the passage of air from a waste pipe into a fixture and possibly into habited spaces.

5. Cross-connections are physical connections between any pipe or receptacle intended to supply water fit for human consumption and any other source of water that is undrinkable for whatever reason. Prevention of cross-connections is obtained by installing an approved air gap, check valve, or backflow preventer.

6. Plumbing fixtures are any installed devices or receptacles intended to dispense potable hot or cold water and discharge that water into a drainage system.

7. Waste systems and pipes are the discharge piping from any fixture, appliance, or appurtenance connected to a drainage system that does not contain fecal matter.

8. Sanitary waste piping and sanitary sewers for plumbing purposes carry the discharge from any fixture that contains fecal matter. For pharmaceutical purposes, the word *sanitary* denotes a system that is superclean and/or sterile.

BASICS PRIOR TO THE TEST

1. Eat a balanced meal prior to the test.
2. Find what, if any, aids, approved references, or books are allowed for the examination.
3. Read the directions for the test, and make sure that you know how long you have.
4. You may want to take water or a snack into the examination room.
5. Find what, if any, pencils or scrap paper is necessary or allowed.
6. Make sure that you understand the test procedures. If not, ask the proctor to explain.
7. Bring two current forms of identification with you.
8. This book is based on the *International Plumbing Code* (*IPC*), primarily the 2012 edition, with new information that has been added to the previous edition of this book.
9. If calculators are allowed, bring additional fresh batteries.
10. Find out if partial credit is given for any problem not completely finished.
11. Find out if there is any penalty for guessing.

DURING THE TEST

1. Scan through the exam quickly. Do the easiest questions first. Do this twice to help select other answers.
2. When answering a question, eliminate the answers that you know are not right.
3. A positive choice is more likely to be correct than a negative one.
4. Read the directions carefully. If there are any questions, ask the proctor.
5. If there are two choices that are direct opposites, one of them is usually correct.
6. Do all the answers appear to be logical? Eliminate those which are not.
7. Do not overcomplicate a problem. Use common sense to clarify anything that is not clear.
8. Read the question and all answers carefully. Do not select the wrong answer because of misreading.
9. Key words and the context clues in each question may help in determining the best answer.
10. Simplicity is a key. You should refrain from selecting a complicated answer if a simple one will do.
11. Added phrases contribute nothing to the meaning of a question, such as *to be* or *of which.*
12. Correct answers usually will not contain any ambiguous words.
13. Select answers that answer the question. Other answers may often contain choices that are wrong.

Appendix

FACTS, FIGURES, CALCULATIONS, AND CONVERSIONS

AVAILABLE LENGTHS OF COPPER PLUMBING TUBE

Drawn (hard copper) (feet)		*Annealed (soft copper) (feet)*	
Type K Tube			
Straight Lengths:		Straight Lengths:	
Up to 8-in. diameter	20	Up to 8-in. diameter	20
10-in. diameter	18	10-in. diameter	18
12-in. diameter	12	12-in. diameter	12
		Coils:	
		Up to 1-in. diameter	60
			100
		1¼-in. diameter	60
			100
		2-in. diameter	40
			45
Type L Tube			
Straight Lengths:		Straight Lengths:	
Up to 10-in. diameter	20	Up to 10-in. diameter	20
12-in. diameter	18	12-inch diameter	18
		Coils:	
		Up to 1-in. diameter	60
		100	
		1¼- and 1½-in. diameter	60
		100	
		2-in. diameter	40
		45	
DWV Tube			
Straight Lengths:		Not available	
All diameters	20		
Type M Tube			
Straight Lengths:		Not available	
All diameters	20		

FIGURE A.1 Available lengths of copper plumbing tube.

COPPER TUBE

Inside Diameter (inches)	*Nominal Size (inches)*	*Outside Diameter (inches)*
Type DWV		
N/A	1/4	0.375
N/A	3/8	0.500
N/A	1/2	0.625
N/A	5/8	0.750
N/A	3/4	0.875
N/A	1	1.125
1.295	1¼	1.375
1.511	1½	1.625
2.041	2	2.125
3.035	2½	2.625
N/A	3	3.125
N/A	3½	3.625
4.009	4	4.125
4.981	5	5.125
5.959	6	6.125
N/A	8	8.125
N/A	10	10.125
N/A	12	12.125

FIGURE A.2 Copper tube.

COPPER TUBE

Nominal Pipe Size (inches)	*Outside Diameter (inches)*	*Inside Diameter (inches)*
Type K		
1/4	0.375	0.305
3/8	0.500	0.402
1/2	0.625	0.527
5/8	0.750	0.652
3/4	0.875	0.745
1	1.125	0.995
1¼	1.375	1.245
1½	1.625	1.481
2	2.125	1.959
2½	2.625	2.435
3	3.125	2.907
3½	3.625	3.385
4	4.125	3.857
5	5.125	4.805
6	6.125	5.741
8	8.125	7.583
10	10.125	9.449
12	12.125	11.315
Type L		
1/4	0.375	0.315
3/8	0.500	0.430
1/2	0.625	0.545
5/8	0.750	0.666
3/4	0.875	0.785
1	1.125	1.025
1¼	1.375	1.265
1½	1.625	1.505
2	2.125	1.985
2½	2.625	2.465
3	3.125	2.945
3½	3.625	3.425
4	4.125	3.905
5	5.125	4.875
6	6.125	5.845
8	8.125	7.725
10	10.125	9.625
12	12.125	11.565

FIGURE A.3 Copper tube *(continued)*.

COPPER TUBE

Nominal Pipe Size (inches)	Outside Diameter (inches)	Inside Diameter (inches)
Type M		
1/4	0.375	0.325
3/8	0.500	0.450
1/2	0.625	0.569
5/8	0.750	0.690
3/4	0.875	0.811
1	1.125	1.055
1¼	1.375	1.291
1½	1.625	1.527
2	2.125	2.009
2½	2.625	2.495
3	3.125	2.981
3½	3.625	3.459
4	4.125	3.935
5	5.125	4.907
6	6.125	5.881
8	8.125	7.785
10	10.125	9.701
12	12.125	11.617

FIGURE A.3 Copper tube *(continued)*.

POLYVINYL CHLORIDE PLASTIC PIPE (PVC)

Nominal Pipe Size (inches)	Outside Diameter (inches)	Inside Diameter (inches)	Wall Thickness (inches)
1/2	0.840	0.750	0.045
3/4	1.050	0.940	0.055
1	1.315	1.195	0.060
1¼	1.660	1.520	0.070
1½	1.900	1.740	0.080
2	2.375	2.175	0.100
2½	2.875	2.635	0.120
3	3.500	3.220	0.140
4	4.500	4.110	0.195

FIGURE A.4 Polyvinyl chloride plastic pipe (PVC).

BRASS PIPE

Nominal Pipe Size (inches)	Outside Diameter (inches)	Inside Diameter (inches)	Wall Thickness (inches)
1/8	0.405	0.281	0.062
1/4	0.376	0.376	0.082
3/8	0.675	0.495	0.090
1/2	0.840	0.626	0.107
3/4	1.050	0.822	0.144
1	1.315	1.063	0.126
1¼	1.660	1.368	0.146
1½	1.900	1.600	0.150
2	2.375	2.063	0.156
2½	2.875	2.501	0.187
3	3.500	3.062	0.219

FIGURE A.5 Brass pipe.

STEAM PIPE EXPANSION (INCHES INCREASE PER 100 IN.)

Temperature (°F)	Steel	Cast Iron	Brass and Copper
0	0	0	0
20	0.15	0.10	0.25
40	0.30	0.25	0.45
60	0.45	0.40	0.65
80	0.60	0.55	0.90
100	0.75	0.70	1.15
120	0.90	0.85	1.40
140	1.10	1.00	1.65
160	1.25	1.15	1.90
180	1.45	1.30	2.15
200	1.60	1.50	2.40
220	1.80	1.65	2.65
240	2.00	1.80	2.90
260	2.15	1.95	3.15
280	2.35	2.15	3.45
300	2.50	2.35	3.75
320	2.70	2.50	4.05
340	2.90	2.70	4.35
360	3.05	2.90	4.65
380	3.25	3.10	4.95
400	3.45	3.30	5.25
420	3.70	3.50	5.60
440	3.95	3.75	5.95
460	4.20	4.00	6.30
480	4.45	4.25	6.65
500	4.70	4.45	7.05

FIGURE A.6 Steam pipe expansion.

BOILING POINT OF WATER AT VARIOUS PRESSURES

Vacuum in Inches of Mercury	Boiling Point	Vacuum in Inches of Mercury	Boiling Point
29	76.62	14	181.82
28	99.93	13	184.61
27	114.22	12	187.21
26	124.77	11	189.75
25	133.22	10	192.19
24	140.31	9	194.50
23	146.45	8	196.73
22	151.87	7	198.87
21	156.75	6	200.96
20	161.19	5	202.25
19	165.24	4	204.85
18	169.00	3	206.70
17	172.51	2	208.50
16	175.80	1	210.25
15	178.91		

FIGURE A.7 Boiling point of water at various pressures.

SIZES OF ORANGEBURG[a] PIPE

Inside Diameter (inches)	Length (feet)
2	5
3	8
4	8
5	5
6	5

[a]A cardboard-type drain pipe.

FIGURE A.8 Sizes of Orangeburg® pipe.

WEIGHT OF CAST-IRON SOIL PIPE

Size (inches)	Service Weight Per Linear Foot (pounds)	Extra Heavy Size (inches)	Per Linear Foot (pounds)
2	4	2	5
3	6	3	9
4	9	4	12
5	12	5	15
6	15	6	19
7	20	8	30
8	25	10	43
		12	54
		15	75

FIGURE A.9 Weight of cast-iron soil pipe.

WEIGHT OF CAST-IRON PIPE

	Diameter (inches)	Service Weight (lb)	Extra Heavy Weight (lb)
Double Hub, 5-ft Lengths	2	21	26
	3	31	47
	4	42	63
	5	54	78
	6	68	100
	8	105	157
	10	150	225
Double Hub, 30-ft Length	2	11	14
	3	17	26
	4	23	33
Single Hub, 5-ft Lengths	2	20	25
	3	30	45
	4	40	60
	5	52	75
	6	65	95
	8	100	150
	10	145	215
Single Hub, 10-ft Lengths	2	38	43
	3	56	83
	4	75	108
	5	98	133
	6	124	160
	8	185	265
	10	270	400
No-Hub Pipe, 10-ft Lengths	1½	27	
	2	38	
	3	54	
	4	74	
	5	95	
	6	118	
	8	180	

FIGURE A.10 Weight of cast-iron pipe.

COMMON SEPTIC TANK CAPACITIES

Single-Family Dwellings; Number of Bedrooms	*Multiple Dwelling Units or Apartments; One Bedroom Each*	*Other Uses; Maximum Fixture-Units Served*	*Minimum Septic Tank Capacity in Gallons*
1–3		20	1000
4	2	25	1200
5–6	3	33	1500
7–8	4	45	2000
	5	55	2250
	6	60	2500
	7	70	2750
	8	80	3000
	9	90	3250
	10	100	3500

FIGURE A.11 Common septic tank capacities.

FACTS ABOUT WATER

1 ft^3 of water contains 7½ gal, 1728 in.^{3}, and weighs 62½ lb.

1 gal of water weighs 8⅓ lb and contains 231 in.^{3}

Water expands 1/23 of its volume when heated from 40° to 212°.

The height of a column of water, equal to a pressure of 1 lb/in.^{2}, is 2.31 ft.

To find the pressure in lb/in.^{2} of a column of water, multiply the height of the column in feet by 0.434.

The average pressure of the atmosphere is estimated at 14.7 lb/in.^{2} so that with a perfect vacuum it will sustain a column of water 34 ft high.

To evaporate 1 ft^3 of water requires the consumption of 7½ lb of ordinary coal or about 1 lb of coal to 1 gal of water.

1 in.^{3} of water evaporated at atmospheric pressure is converted into approximately 1 ft^3 of steam.

FIGURE A.12 Facts about water.

RATES OF WATER FLOW

Fixture	*Flow Rate (gpm)*[a]
Ordinary basin faucet	2.0
Self-closing basin faucet	2.5
Sink faucet, 3/8 in.	4.5
Sink faucet, 1/2 in.	4.5
Bathtub faucet	6.0
Laundry tub, 1/2 in.	5.0
Shower	5.0
Flush tank for the water closet	3.0
Flushometer valve for water closet	5.0
Flushometer valve for urinal	2.0
Drinking fountain	.75
Sillcock (wall hydrant)	5.0

[a]Figures do not represent the use of water-conservation devices.

FIGURE A.13 Rates of water flow.

CONSERVING WATER

Activity	*Normal Use (gallons)*	*Conservative Use (gallons)*
Shower	6 (water running)	4 (wet down, soap up, rinse off)
Tub bath	15 (full)	10–12 (minimal water level)
Dishwashing	20 (tap running)	5 (wash and rinse in sink)
Toilet flushing	5–7 (depends on tank size)	1½–3
Automatic dishwasher	16 (full cycle)	7 (short cycle)
Washing machine	60 (full cycle, top water level)	27 (short cycle, minimal water level)
Washing hands	2 (tap running)	1 (full basin)
Brushing teeth	1 (tap running)	1/2 (wet and rinse briefly)

FIGURE A.14 Conserving water.

DEMAND FOR WATER AT INDIVIDUAL OUTLETS

Type of Outlet	*Demand (gpm)*
Ordinary lavatory faucet	2.0
Self-closing lavatory faucet	2.0
Sink faucet, 3/8 in. or 1/2 in.	2.5
Sink faucet, 3/4 in.	4.0
Bath faucet, 1/2 in.	5.0
Shower head, 1/2 in.	5.0
Laundry faucet, 1/2 in.	5.0
Water closet flush tank	3.0
Flush valve (25 psi flow pressure)	25.0
Flush valve (15 psi flow pressure)	20.0
1/2-in. flush valve (15 psi flow pressure)	15.0
Drinking fountain jet	0.75
Dishwashing machine (domestic)	4.0
Laundry machine (8 or 16 lb)	4.0
Aspirator (operating room or laboratory)	2.5
Hose bib or sillcock (1/2 in.)	5.0

FIGURE A.15 Demand for water at individual outlets.

MAXIMUM DISTANCES FROM FIXTURE TRAPS TO VENTS

Size of Fixture (inches)	*Distance from Trap to Vent*[a]
1¼	2 ft 6 in.
1½	3 ft 6 in.
2	5 ft
3	6 ft
4	10 ft

[a]Figures may vary with local plumbing codes.

FIGURE A.16 Maximum distances from fixture traps to vents.

BRANCH PIPING FOR FIXTURES—MINIMUM SIZE

Fixture or Device	*Size (inches)*
Bathtub	1/2
Combination sink and laundry tray	1/2
Drinking fountain	3/8
Dishwashing machine (domestic)	1/2
Kitchen sink (domestic)	1/2
Kitchen sink (commercial)	3/4
Lavatory	3/8
Laundry tray (1, 2, or 3 compartments)	1/2
Shower (single head)	1/2
Sink (service, slop)	1/2
Sink (flushing rim)	3/4
Urinal (1-in. flush valve)	1
Urinal (3/4-in. flush valve)	3/4
Urinal (flush tank)	1/2
Water closet (flush tank)	3/8
Water closet (flush valve)	1
Hose bib	1/2
Wall hydrant or sillcock	1/2

FIGURE A.17 Branch piping for fixtures—minimum size.

POTENTIAL SEWAGE FLOWS

Type of Establishment	*Gallons (per day per person)*[a]
Schools (toilets and lavatories only)	15
Schools (with above plus cafeteria)	25
Schools (with above plus cafeteria and showers)	35
Day workers at schools and offices	15
Day camps	25
Trailer parks or tourist camps (with built-in bath)	50
Trailer parks or tourist camps (with central bathhouse)	35
Work or construction camps	50
Public picnic parks (toilet wastes only)	5
Public picnic parks (bathhouse, showers, and flush toilets)	10
Swimming pools and beaches	10
Country clubs	25 per locker
Luxury residences and estates	150
Rooming houses	40
Boarding houses	50
Hotels (with connecting baths)	50
Hotels (with private baths, 2 persons per room)	100
Boarding schools	100
Factories (gallons per person per shift, exclusive of industrial wastes)	25
Nursing homes	75
General hospitals	150
Public institutions (other than hospitals)	100
Restaurants (toilet and kitchen wastes per unit of serving capacity)	25
Kitchen wastes from hotels, camps, boarding houses, etc. that serve 3 meals per day	10
Motels	50 per bed

[a]Except for country clubs and motels.

FIGURE A.18 Potential sewage flows.

POTENTIAL SEWAGE FLOWS

Type of Establishment	*Gallons*
Motels with bath, toilet and kitchen wastes	60 per bed space
Drive-in theaters	5 per car space
Stores	400 per toilet room
Service stations	10 per vehicle served
Airports	3–5 per passenger
Assembly halls	2 per person
Bowling alleys	75 per lane
Churches (small)	3–5 per seat
Churches (large with kitchens)	5–7 per seat
Dance halls	2 per day per person
Laundries (coin-operated)	400 per machine
Service stations	10 per car
Subdivisions or individual homes	75 per day per person
Marinas:	
Flush toilets	36 per fixture per hr
Urinals	10 per fixture per hr
Wash basins	15 per fixture per hr
Showers	150 per fixture per hr

FIGURE A.19 Potential sewage flows.

R VALUES FOR COMMON INSULATING MATERIALS

Material	*R Value per Inch*
Perlite	2.75
Loose-fill mineral wool	3.70
Extruded polystyrene	5.0
Urethane	7.2–8.0
Urethane w/foil face and 3/4-in. air space	10.0
Fiberglass batt	3.17

FIGURE A.20 R values for common insulating materials.

MELTING POINT OF VARIOUS MATERIALS

Material	*Degrees Fahrenheit*
Aluminum	1218
Antimony	1150
Brass	1800
Bronze	1700
Chromium	2740
Copper	2450
Gold	1975
Iron (cast)	2450
Iron (wrought)	2900
Lead	620
Manganese	2200
Mercury	39.5
Molybdenum	4500
Monel	2480
Platinum	3200
Steel (mild)	2600
Steel (stainless)	2750
Tin	450
Titanium	3360
Zinc	787

FIGURE A.21 Melting point of various materials.

WEIGHTS OF VARIOUS MATERIALS

Material	*Pounds per Cubic Inch*	*Pounds per Cubic Foot*
Aluminum	0.093	160
Antimony	0.2422	418
Brass	0.303	524
Bronze	0.320	552
Chromium	0.2348	406
Copper	0.323	558
Gold	0.6975	1205
Iron (cast)	0.260	450
Iron (wrought)	0.2834	490
Lead	0.4105	710
Manganese	0.2679	463
Mercury	0.491	849
Molybdenum	0.309	534
Monel	0.318	550
Platinum	0.818	1413
Steel (mild)	0.2816	490
Steel (stainless)	0.277	484
Tin	0.265	459
Titanium	0.1278	221
Zinc	0.258	446

FIGURE A.22 Weights of various materials.

MELTING POINTS OF COMMERCIAL METALS

Metal	*Degrees Fahrenheit*
Aluminum	1200
Antimony	1150
Bismuth	500
Brass	1700/1850
Copper	1940
Cadmium	610
Iron (cast)	2300
Iron (wrought)	2900
Lead	620
Mercury	139
Steel	2500
Tin	446
Zinc, cast	785

FIGURE A.23 Melting points of commercial metals.

Pipe size (in inches)	PSI	Length of pipe is 50 feet
¾	20	16
¾	40	24
¾	60	29
¾	80	34
1	20	31
1	40	44
1	60	55
1	80	65
1¼	20	84
1¼	40	121
1¼	60	151
1¼	80	177
1½	20	94
1½	40	137
1½	60	170
1½	80	200

FIGURE A.24 Discharge of pipes in gallons per minute.

Pipe size (in inches)	PSI	Length of pipe is 100 feet
¾	20	11
¾	40	16
¾	60	20
¾	80	24
1	20	21
1	40	31
1	60	38
1	80	44
1¼	20	58
1¼	40	84
1¼	60	104
1¼	80	121
1½	20	65
1½	40	94
1½	60	117
1½	80	137

FIGURE A.25 Discharge of pipes in gallons per minute.

Hose size (in inches)	Maximum outside diameter	Threads per inch
¼	1.0625	11.5

FIGURE A.26 Threads per inch for garden hose.

Valve size (in inches)	Number of turns required to operate valve
3	7.5
4	14.5
6	20.5
8	27
10	33.5

FIGURE A.27 The number of turns required to operate a double-disk valve.

Valve size (in inches)	Number of turns required to operate valve
3	11
4	14
6	20
8	27
10	33

FIGURE A.28 Number of turns required to operate a metal-seated sewerage valve.

Pipe diameter (in inches)	Approximate capacity (in U.S. gallons) per foot of pipe
¾	.0230
1	.0408
1¼	.0638
1½	.0918
2	.1632
3	.3673
4	.6528
6	1.469
8	2.611
10	4.018

FIGURE A.29 Pipe capacities.

	Temperature Change (°F)						
Length (ft)	40	50	60	70	80	90	100
20	0.278	0.348	0.418	0.487	0.557	0.626	0.696
40	0.557	0.696	0.835	0.974	1.114	1.235	1.392
60	0.835	1.044	1.253	1.462	1.670	1.879	2.088
80	1.134	1.392	1.670	1.879	2.227	2.506	2.784
100	1.192	1.740	2.088	2.436	2.784	3.132	3.480

FIGURE A.30 Thermal expansion of PVE-DVW.

	Coefficient	
Pipe material	in/in/°F	(°C)
Metallic pipe		
Carbon steel	0.000005	(14.0)
Stainless steel	0.000115	(69)
Cast iron	0.0000056	(1.0)
Copper	0.000010	(1.8)
Aluminum	0.0000980	(1.7)
Brass (yellow)	0.000001	(1.8)
Brass (red)	0.000009	(1.4)
Plastic pipe		
ABS	0.00005	(8)
PVC	0.000060	(33)
PB	0.000150	(72)
PE	0.000080	(14.4)
CPVC	0.000035	(6.3)
Styrene	0.000060	(33)
PVDF	0.000085	(14.5)
PP	0.000065	(77)
Saran	0.000038	(6.5)
CAB	0.000080	(14.4)
FRP (average)	0.000011	(1.9)
PVDF	0.000096	(15.1)
CAB	0.000085	(14.5)
HDPE	0.00011	(68)
Glass		
Borosilicate	0.0000018	(0.33)

FIGURE A.31 Thermal expansion of piping materials.

Length (ft)	Temperature Change (°F) 40	50	60	70	80	90	100
20	0.536	0.670	0.804	0.938	1.072	1.206	1.340
40	1.070	1.340	1.610	1.880	2.050	2.420	2.690
60	1.609	2.010	2.410	2.820	3.220	3.620	4.020
80	2.143	2.680	3.220	3.760	4.290	4.830	5.360
100	2.680	3.350	4.020	4.700	5.360	6.030	6.700

FIGURE A.32 Thermal expansion of all pipes (except PVE-DWV).

Pipe size (in inches)	Maximum outside diameter	Threads per inch
¼	1.375	8
1	1.375	8
1¼	1.6718	9
1½	1.990	9
2	2.5156	8
3	3.6239	6
4	5.0109	4
5	6.260	4
6	7.025	4

FIGURE A.33 Threads per inch for national standard thread.

Pipe size (in inches)	Maximum outside diameter	Threads per inch
¼	1.0353	14
1	1.295	11.5
1¼	1.6399	11.5
1½	1.8788	11.5
2	2.5156	8
3	3.470	8
4	4.470	8

FIGURE A.34 Threads per inch for American Standard Straight Pipe.

IPS, in	Weight per foot, lb	Length in feet containing 1 ft³ of water	Gallons in 1 linear ft
¼	0.42		0.005
⅜	0.57	754	0.0099
½	0.85	473	0.016
¾	1.13	270	0.027
1	1.67	166	0.05
1¼	2.27	96	0.07
1½	2.71	70	0.1
2	3.65	42	0.17
2½	5.8	30	0.24
3	7.5	20	0.38
4	10.8	11	0.66
5	14.6	7	1.03
6	19.0	5	1.5
8	25.5	3	2.6
10	40.5	1.8	4.1
12	53.5	1.2	5.9

FIGURE A.35 Weight of steel pipe and contained water.

Pipe	Weight factor*
Aluminum	0.35
Brass	1.12
Cast iron	0.91
Copper	1.14
Stainless steel	1.0
Carbon steel	1.0
Wrought iron	0.98

*Average plastic pipe weights one-fifth as much as carbon steel pipe.

FIGURE A.36 Relative weight factor for metal pipe.

Pipe size, in	Rod size, in
2 and smaller	⅜
2½ to 3½	½
4 and 5	⅝
6	¾
8 to 12	⅞
14 and 16	1
18	1⅛
20	1¼
24	1½

FIGURE A.37 Recommended rod size for individual pipes.

Nominal rod diameter, in	Root area of thread, in^2	Maximum safe load at rod temperature of 650°F, lb
¼	0.027	240
5/16	0.046	410
⅜	0.068	610
½	0.126	1,130
⅝	0.202	1,810
¾	0.302	2,710
⅞	0.419	3,770
1	0.552	4,960
1⅛	0.693	6,230
1¼	0.889	8,000
1⅜	1.053	9,470
1½	1.293	11,630
1⅝	1.515	13,630
1¾	1.744	15,690
1⅞	2.048	18,430
2	2.292	20,690
2¼	3.021	27,200
2½	3.716	33,500
2¾	4.619	41,600
3	5.621	50,600
3¼	6.720	60,500
3½	7.918	71,260

FIGURE A.38 Load rating of threaded rods.

	No water pressure	No water pressure at fixture	Low water pressure to fixture
Street water main	X		X
Curb stop	X		X
Water service	X		X
Branches		X	X
Valves	X	X	X
Stems, washers (hot and cold)		X	X
Aerator		X	X
Water meter	X	X	X

FIGURE A.39 Where to look for causes of water-pressure problems.

Nominal pipe size in inches	Outside diameter in inches	Inside diameter in inches	Weight per linear foot in pounds
¼	.375	.305	.145
⅜	.500	.402	.269
½	.625	.527	.344
⅝	.750	.652	.418
¾	.875	.745	.641
1	1.125	.995	.839
1¼	1.375	1.245	1.040
1½	1.625	1.481	1.360
2	2.125	1.959	2.060
2½	2.625	2.435	2.932
3	3.125	2.907	4.000
3½	3.625	3.385	5.122
4	4.125	3.857	6.511
5	5.125	4.805	9.672
6	6.125	5.741	13.912
8	8.125	7.583	25.900
10	10.125	9.449	40.322
12	12.125	11.315	57.802

FIGURE A.40 Facts about copper Type K tubing.

Nominal pipe size in inches	Outside diameter in inches	Inside diameter in inches	Weight per linear foot in pounds
¼	.375	.315	.126
⅜	.500	.430	.198
½	.625	.545	.285
⅝	.750	.666	.362
¾	.875	.785	.455
1	1.125	1.025	.655
1¼	1.375	1.265	.884
1½	1.625	1.505	1.111
2	2.125	1.985	1.750
2½	2.625	2.465	2.480
3	3.125	2.945	3.333
3½	3.625	3.425	4.290
4	4.125	3.905	5.382
5	5.125	4.875	7.611
6	6.125	5.845	10.201
8	8.125	7.725	19.301
10	10.125	9.625	30.060
12	12.125	11.565	40.390

FIGURE A.41 Facts about copper Type L tubing.

Nominal pipe size in inches	Outside diameter in inches	Inside diameter in inches	Weight per linear foot in pounds
¼	.375	.325	.107
⅜	.500	.450	.145
½	.625	.569	1.204
⅝	.750	.690	.263
¾	.875	.811	.328
1	1.125	1.055	.465
1¼	1.375	1.291	.682
1½	1.625	1.527	.940
2	2.125	2.009	1.460
2½	2.625	2.495	2.030
3	3.125	2.981	2.680
3½	3.625	3.459	3.580
4	4.125	3.935	4.660
5	5.125	4.907	6.660
6	6.125	5.881	8.920
8	8.125	7.785	16.480
10	10.125	9.701	25.590
12	12.125	11.617	36.710

FIGURE A.42 Facts about copper Type M tubing.

Diameter	Service weight	Extra heavy weight
2"	11	14
3"	17	26
4"	23	33

FIGURE A.43 Weight of double-hub cast-iron pipe (30-inch length).

Diameter	Service weight	Extra heavy weight
2"	38	43
3"	56	83
4"	75	108
5"	98	133
6"	124	160
8"	185	265
10"	270	400

FIGURE A.44 Weight of single-hub cast-iron pipe (10-foot length).

Size	Lbs. per ft.
2"	4
3"	6
4"	9
5"	12
6"	15
7"	20
8"	25

FIGURE A.45 Weight of cast-iron soil pipe.

Diameter	Service weight	Extra heavy weight
2"	21	26
3"	31	47
4"	42	63
5"	54	78
6"	68	100
8"	105	157
10"	150	225

FIGURE A.46 Weight of double-hub cast-iron pipe (5-foot length).

Diameter	Service weight	Extra heavy weight
2"	20	25
3"	30	45
4"	40	60
5"	52	75
6"	65	95
8"	100	150
10"	145	215

FIGURE A.47 Weight of single-hub cast-iron pipe (5-foot length).

Diameter	Weight
1½"	27
2"	38
3"	54
4"	74
5"	95
6"	118
8"	180

FIGURE A.48 Weight of no-hub cast-iron pipe (10-foot length).

Diameter	Weight
2"	5
3"	9
4"	12
5"	15
6"	19
8"	30
10"	43
12"	54
15"	75

FIGURE A.49 Weight of extra-heavy cast-iron soil pipe.

Quantity	Equals
1 square meter	10.764 square feet 1.196 square yards
1 square centimeter	.155 square inch
1 square millimeter	.00155 square inch
.836 square meter	1 square yard
.0929 square meter	1 square foot
6.452 square centimeter	1 square inch
645.2 square millimeter	1 square inch

FIGURE A.50 Surface measures.

Quantity	Equals
144 sq. inches	1 sq. foot
9 sq. feet	1 sq. yard
1 sq. yard	1296 sq. inches
4840 sq. yards	1 acre
640 acres	1 sq. mile

FIGURE A.51 Square measure.

Quantity	Equals	Equals
7.92 inches	1 link	
100 links	1 chain	66 feet
10 chains	1 furling	660 feet
80 chains	1 mile	5280 feet

FIGURE A.52 Surveyor's measure.

Inches	Millimeters
1	25.4
2	50.8
3	76.2
4	101.6
5	127.0
6	152.4
7	177.8
8	203.2
9	228.6
10	254.0
11	279.4
12	304.8
13	330.2
14	355.6
15	381.0
16	406.4
17	431.8
18	457.2
19	482.6
20	508.0

FIGURE A.53 Inches to millimeters.

Inches^2	Millimeters^2
0.01227	8.0
0.04909	31.7
0.11045	71.3
0.19635	126.7
0.44179	285.0
0.7854	506.7
1.2272	791.7
1.7671	1140.1
3.1416	2026.8
4.9087	3166.9
7.0686	4560.4
12.566	8107.1
19.635	12667.7
28.274	18241.3
38.485	24828.9
50.265	32478.9
63.617	41043.1
78.540	50670.9

FIGURE A.54 Area in inches and millimeters.

Inches	DN
0.3927	10
0.7854	20
1.1781	30
1.5708	40
2.3562	60
3.1416	80
3.9270	100
4.7124	120
6.2832	160
7.8540	200
9.4248	240
12.566	320
15.708	400
18.850	480
21.991	560
25.133	640
28.274	720
31.416	800

FIGURE A.55 Circumference in inches and SI units.

Quantity	Unit	Symbol
Time	Second	s
Plane angle	Radius	rad
Force	Newton	N
Energy, work, quantity of heat	Joule	J
	Kilojoule	kJ
	Megajoule	MJ
Power, heat flow rate	Watt	W
	Kilowatt	kW
Pressure	Pascal	Pa
	Kilopascal	kPa
	Megapascal	MPa
Velocity, speed	Meter per second	m/s
	Kilometer per hour	km/h

FIGURE A.56 Metric symbols.

Units	Equals
1 decimeter	4 inches
1 meter	1.1 yards
1 kilometer	⅝ mile
1 hektar	2½ acres
1 stere or cu. meter	¼ cord
1 liter	1.06 qt. liquid; 0.9 qt. dry
1 hektoliter	2⅝ bushel
1 kilogram	2⅕ lbs.
1 metric ton	2200 lbs.

FIGURE A.57 Approximate metric equivalents.

Feet	Meters (m)	Millimeters (mm)
1	0.305	304.8
2	0.610	609.6
3 (1 yd.)	0.914	914.4
4	1.219	1 219.2
5	1.524	1 524.0
6 (2 yd.)	1.829	1 828.8
7	2.134	2 133.6
8	2.438	2 438.2
9 (3 yd.)	2.743	2 743.2
10	3.048	3 048.0
20	6.096	6 096.0
30 (10 yd.)	9.144	9 144.0
40	12.19	12 192.0
50	15.24	15 240.0
60 (20 yd.)	18.29	18 288.0
70	21.34	21 336.0
80	24.38	24 384.0
90 (30 yd.)	27.43	27 432.0
100	30.48	30 480.0

FIGURE A.58 Length conversions.

Quantity	Unit	Symbol
Length	Millimeter	mm
	Centimeter	cm
	Meter	m
	Kilometer	km
Area	Square Millimeter	mm^7
	Square Centimeter	cm^2
	Square Decimeter	dm^2
	Square Meter	m^2
	Square Kilometer	km^2
Volume	Cubic Centimeter	cm^3
	Cubic Decimeter	dm^3
	Cubic Meter	m^3
Mass	Milligram	mg
	Gram	g
	Kilogram	kg
	Tonne	t
Temperature	Degree Celsius	°C
	Kelvin	K
Time	Second	s
Plane angle	Radius	rad
Force	Newton	N
Energy, work, quantity of heat	Joule	J
	Kilojoule	kJ
	Megajoule	MJ
Power, heat flow rate	Watt	W
	Kilowatt	kW
Pressure	Pascal	Pa
	Kilopascal	kPa
	Megapascal	MPa

FIGURE A.59 Metric symbols.

Square feet	Square meters
1	0.925
2	.1850
3	.2775
4	.3700
5	.4650
6	.5550
7	.6475
8	.7400
9	.8325
10	.9250
25	2.315
50	4.65
100	9.25

FIGURE A.60 Square feet to square meters.

Quantity	Equals
Metric cubic measure	
1000 cubic millimeters (cu. mm.)	1 cubic centimeter
1000 cubic centimeters (cu. cm.)	1 cubic decimeter
1000 cubic decimeters (cu. dm.)	1 cubic meter
Metric capacity measure	
10 milliliters (mi.)	1 centiliter
10 centiliters (cl.)	1 deciliter
10 deciliters (dl.)	1 liter
10 liters (l.)	1 dekaliter
10 dekaliters (Dl.)	1 hectoliter
10 hectoliters (hl.)	1 kiloliter
10 kiloliters (kl.)	1 myrialiter (ml.)

FIGURE A.61 Metric cubic measure.

Quantity	Equals
10 milligrams (mg.)	1 centigram
10 centigrams (cg.)	1 decigram
10 decigrams (dg.)	1 gram
10 grams (g.)	1 dekagram
10 dekagrams (Dg.)	1 hectogram
10 hectograms (hg.)	1 kilogram
10 kilogram (kg.)	1 myriagram
10 myriagrams (Mg.)	1 quintal

FIGURE A.62 Metric weight measure.

Metric linear measure		
Measure	**Equals**	**Equals**
	1 millimeter	.001 meter
10 millimeter	1 centimeter	.01 meter
10 centimeter	1 decimeter	.1 meter
10 decimeter	1 meter	1 meter
10 meters	1 dekameter	10 meters
10 dekameters	1 hectometer	100 meters
10 hectometers	1 kilometer	1000 meters
10 kilometers	1 myriameter	10,000 meters

Metric land measure	
Unit	**Equals**
1 centiare (ca.)	1 sq. meter
100 centiares (ca.)	1 are
100 ares (A.)	1 hectare
100 hectares (ha.)	1 sq. kilometer

FIGURE A.63 Metric linear measure.

1 cu. ft. at 50°F. weighs 62.41 lb.
1 gal. at 50°F weighs 8.34 lb.
1 cu. ft. of ice weighs 57.2 lb.
Water is at its greatest density at 39.2°F.
1 cu. ft. at 39.2°F. weighs 62.43 lb.

FIGURE A.64 Water weight.

Quantity	Equals	Equals
12 inches	1 foot	
3 feet	1 yard	36 inches
5½ yards	1 rod	16½ feet
40 rods	1 furlong	660 feet
8 furlongs	1 mile	5280 feet

FIGURE A.65 Linear measure.

Quantity	Equals
Linear measure	
12 inches	1 foot
3 feet	1 yard
5½ yards	1 rod
320 rods	1 mile
1 mile	1760 yards
1 mile	5280 feet
Square measure	
144 sq. inches	1 sq. foot
9 sq. feet	1 sq. yard
1 sq. yard	1296 sq. inches
4840 sq. yards	1 acre
640 acres	1 sq. mile

FIGURE A.66 Weights and measures.

Quantity	Equals
1 meter	39.3 inches 3.28083 feet 1.0936 yards
1 centimeter	.3937 inch
1 millimeter	.03937 inch, or nearly ⅟₂₅ inch
1 kilometer	0.62137 mile
.3048 meter	1 foot
2.54 centimeters	1 inch 25.40 millimeters

FIGURE A.67 Metric conversions.

Quantity	Equals
100 sq. millimeters	1 sq. centimeter
100 sq. centimeters	1 sq. decimeter
100 sq. decimeters	1 sq. meter

FIGURE A.68 Metric square measure.

Unit	Equals
1 gallon	0.133681 cubic foot
1 gallon	231 cubic inches

FIGURE A.69 Volume measure equivalent.

To change	To	Multiply by
Inches	Millimeters	25.4
Feet	Meters	.3048
Miles	Kilometers	1.6093
Square inches	Square centimeters	6.4515
Square feet	Square meters	.09290
Acres	Hectares	.4047
Acres	Square kilometers	.00405
Cubic inches	Cubic centimeters	16.3872
Cubic feet	Cubic meters	.02832
Cubic yards	Cubic meters	.76452
Cubic inches	Liters	.01639
U.S. gallons	Liters	3.7854
Ounces (avoirdupois)	Grams	28.35
Pounds	Kilograms	.4536
Lbs. per sq. in. (P.S.I.)	Kg.'s per sq. cm.	.0703
Lbs. per cu. ft.	Kg.'s per cu. meter	16.0189
Tons (2000 lbs.)	Metric tons (1000 kg.)	.9072
Horsepower	Kilowatts	.746

FIGURE A.70 English to metric conversions.

Quantity	Equals
Cubic measure	
1728 cubic inches	1 cubic foot
27 cubic feet	1 cubic yard
Avoirdupois weight	
16 ounces	1 pound
100 pounds	1 hundredweight
20 hundredweight	1 ton
1 ton	2000 pounds
1 long ton	2240

FIGURE A.71 Weights and measures.

Unit	Equals
1 cu. ft.	62.4 lbs.
1 cu. ft.	7.48 gal.
1 gal.	8.33 lbs.
1 gal.	0.1337 cu. ft.

FIGURE A.72 Water volume to weight conversion.

Unit	Equals
1 meter	39.3 inches 3.28083 feet 1.0936 yards
1 centimeter	.3937 inch
1 millimeter	.03937 inch, or nearly 1/25 inch
1 kilometer	0.62137 mile
1 foot	.3048 meter
1 inch	2.54 centimeters 25.40 millimeters
1 square meter	10.764 square feet 1.196 square yards
1 square centimeter	.155 square inch
1 square millimeter	.00155 square inch
1 square yard	.836 square meter
1 square foot	.0929 square meter
1 square inch	6.452 square centimeter 645.2 square millimeter

FIGURE A.73 Metric-customary equivalents.

Quantity	Equals
4 gills	1 pint
2 pints	1 quart
4 quarts	1 gallon
31½ gallons	1 barrel
1 gallon	231 cubic inches
7.48 gallons	1 cubic foot
1 gallon water	8.33 pounds
1 gallon gasoline	5.84 pounds

FIGURE A.74 Liquid measure.

Quantity	Equals
12 inches (in. or ")	1 foot (ft. or ')
3 feet	1 yard (yd.)
5½ yards or 16½ feet	1 rod (rd.)
40 rods	1 furlong (fur.)
8 furlongs or 320 rods	1 statute mile (mi.)

FIGURE A.75 Long measure.

Unit	Equals
1 sq. centimeter	0.1550 sq. in.
1 sq. decimeter	0.1076 sq. ft.
1 sq. meter	1.196 sq. yd.
1 are	3.954 sq. rd.
1 hektar	2.47 acres
1 sq. kilometer	0.386 sq. mile
1 sq. in.	6.452 sq. centimeters
1 sq. ft.	9.2903 sq. decimeters
1 sq. yd.	0.8361 sq. meter
1 sq. rd.	0.2529 are
1 acre	0.4047 hektar
1 sq. mile	2.59 sq. kilometers

FIGURE A.76 Square measure.

Volume	Weight
1 cu. ft. sand	Approx. 100 lbs.
1 cu. yd.	2700 lbs.
1 ton	¾ yd. or 20 cu. ft.
Avg. shovelful	15 lbs.
12 qt. pail	40 lbs.

FIGURE A.77 Sand volume to weight conversion.

U.S.	Metric
0.001 inch	0.025 mm
1 inch	25.400 mm
1 foot	30.48 cm
1 foot	0.3048 m
1 yard	0.9144 m
1 mile	1.609 km

FIGURE A.78 Conversion table.

Unit	Equals
1 cubic meter	35.314 cubic feet 1.308 cubic yards 264.2 U.S. gallons (231 cubic inches)
1 cubic decimeter	61.0230 cubic inches .0353 cubic feet
1 cubic centimeter	.061 cubuic inch
1 liter	1 cubic decimeter 61.0230 cubic inches 0.0353 cubic foot 1.0567 quarts (U.S.) 0.2642 gallon (U.S.) 2.2020 lb. of water at 62°F.
1 cubic yard	.7645 cubic meter
1 cubic foot	.02832 cubic meter 28.317 cubic decimeters 28.317 liters
1 cubic inch	16.383 cubic centimeters
1 gallon (British)	4.543 liters
1 gallon (U.S.)	3.785 liters
1 gram	15.432 grains
1 kilogram	2.2046 pounds
1 metric ton	.9842 ton of 2240 pounds 19.68 cwts. 2204.6 pounds
1 grain	.0648 gram
1 ounce avoirdupois	28.35 grams
1 pound	.4536 kilograms
1 ton of 2240 lb.	1.1016 metric tons 1016 kilograms

FIGURE A.79 Measure of volume and capacity.

U.S.	Metric
1 inch^2	6.4516 cm^2
1 feet^2	0.0929 m^2
1 yard^2	0.8361 m^2
1 acre	0.4047 ha
1 mile^2	2.590 km^2
1 inch^3	16.387 cm^3
1 feet^3	0.0283 m^3
1 yard^3	0.7647 m^3
1 U.S. ounce	29.57 ml
1 U.S. pint	0.4732 l
1 U.S. gallon	3.785 l
1 ounce	28.35 g
1 pound	0.4536 kg

FIGURE A.80 Conversion table.

	Imperial	Metric
Length	1 inch	25.4 mm
	1 foot	0.3048 m
	1 yard	0.9144 m
	1 mile	1.609 km
Mass	1 pound	0.454 kg
	1 U.S. short ton	0.9072 tonne
Area	1 ft^2	0.092 m^2
	1 yd^2	0.836 m^2
	1 acre	0.404 hectare (ha)
Capacity/Volume	1 ft^3	0.028 m^3
	1 yd^3	0.764 m^3
	1 liquid quart	0.946 litre (1)
	1 gallon	3.785 litre (1)
Heat	1 BTU	1055 joule (J)
	1 BTU/hr	0.293 watt (W)

FIGURE A.81 Measurement conversions: Imperial to metric.

To find	Multiply	By
Microns	Mils	25.4
Centimeters	Inches	2.54
Meters	Feet	0.3048
Meters	Yards	0.19144
Kilometers	Miles	1.609344
Grams	Ounces	28.349523
Kilograms	Pounds	0.4539237
Liters	Gallons (U.S.)	3.7854118
Liters	Gallons (Imperial)	4.546090
Milliliters (cc)	Fluid ounces	29.573530
Milliliters (cc)	Cubic inches	16.387064
Square centimeters	Square inches	6.4516
Square meters	Square feet	0.09290304
Square meters	Square yards	0.83612736
Cubic meters	Cubic feet	2.8316847×10^{-2}
Cubic meters	Cubic yards	0.76455486
Joules	BTU	1054.3504
Joules	Foot-pounds	1.35582
Kilowatts	BTU per minute	0.01757251
Kilowatts	Foot-pounds per minute	2.2597×10^{-5}
Kilowatts	Horsepower	0.7457
Radians	Degrees	0.017453293
Watts	BTU per minute	17.5725

FIGURE A.82 Conversion factors in converting from customary (U.S.) units to metric units.

To change	To	Multiply by
Inches	Feet	0.0833
Inches	Millimeters	25.4
Feet	Inches	12
Feet	Yards	0.3333
Yards	Feet	3
Square inches	Square feet	0.00694
Square feet	Square inches	144
Square feet	Square yards	0.11111
Square yards	Square feet	9
Cubic inches	Cubic feet	0.00058
Cubic feet	Cubic inches	1728
Cubic feet	Cubic yards	0.03703
Gallons	Cubic inches	231
Gallons	Cubic feet	0.1337
Gallons	Pounds of water	8.33
Pounds of water	Gallons	0.12004
Ounces	Pounds	0.0625
Pounds	Ounces	16
Inches of water	Pounds per square inch	0.0361
Inches of water	Inches of mercury	0.0735
Inches of water	Ounces per square inch	0.578
Inches of water	Pounds per square foot	5.2
Inches of mercury	Inches of water	13.6
Inches of mercury	Feet of water	1.1333
Inches of mercury	Pounds per square inch	0.4914
Ounces per square inch	Inches of mercury	0.127
Ounces per square inch	Inches of water	1.733
Pounds per square inch	Inches of water	27.72
Pounds per square inch	Feet of water	2.310
Pounds per square inch	Inches of mercury	2.04
Pounds per square inch	Atmospheres	0.0681
Feet of water	Pounds per square inch	0.434
Feet of water	Pounds per square foot	62.5
Feet of water	Inches of mercury	0.8824

FIGURE A.83 Measurement conversion factors.

To change	To	Multiply by
Atmospheres	Pounds per square inch	14.696
Atmospheres	Inches of mercury	29.92
Atmospheres	Feet of water	34
Long tons	Pounds	2240
Short tons	Pounds	2000
Short tons	Long tons	0.89295

FIGURE A.84 Measurement conversion factors.

Quantity	Equals	Cubic inches
2 pints	1 quart	67.2
8 quarts	1 peck	537.61
4 pecks	1 bushel	2150.42

FIGURE A.85 Dry measure.

Unit	Equals
1 gram	15.432 grains
1 kilogram	2.2046 pounds
1 metric ton	.9842 ton of 2240 pounds 19.68 cwts. 2204.6 pounds
1 grain	.0648 gram
1 ounce avoirdupois	28.35 grams
1 pound	.4536 kilograms
1 ton of 2240 lb.	1.1016 metric tons 1016 kilograms

FIGURE A.86 Weight conversions.

Inches	Millimeters
3	76.2
4	101.6
5	127
6	152.4
7	177.8
8	203.2
9	228.6
10	254

FIGURE A.87 Diameter in inches and millimeters.

Quantity	Equals	Equals
144 sq. inches	1 sq. foot	
9 sq. feeet	1 sq. yard	
30¼ sq. yards	1 sq. rod	272.25 sq. feet
160 sq. rods	1 acre	4840 sq. yards or 43,560 sq. feet
640 acres	1 sq. mile	3,097,600 sq. yards
36 sq. miles	1 township	

FIGURE A.88 Square measure.

Quantity	Equals	Meters	English equivalent
1 mm.	1 millimeter	1/1000	.03937 in.
10 mm.	1 centimeter	1/100	.3937 in.
10 cm.	1 decimeter	1/10	3.937 in.
10 dm.	1 meter	1	39.37 in.
10 m.	1 dekameter	10	32.8 ft.
10 Dm.	1hectometer	100	328.09 ft.
10 Hm.	1 kilometer	1000	.62137 mile

FIGURE A.89 Lengths.

Barometer (ins. of mercury)	Pressure (lbs. per sq. in.)
28.00	13.75
28.25	13.88
28.50	14.00
28.75	14.12
29.00	14.24
29.25	14.37
29.50	14.49
29.75	14.61
29.921	14.696
30.00	14.74
30.25	14.86
30.50	14.98
30.75	15.10
31.00	15.23

Rule: Barometer in inches of mercury × .49116 = lbs. per sq. in.

FIGURE A.90 Atmospheric pressure per square inch.

Pipe size	Projected flow rate (gallons per minute)
½ inch	2 to 5
¾ inch	5 to 10
1 inch	10 to 20
1¼ inch	20 to 30
1½ inch	30 to 40

FIGURE A.91 Projected flow rates for various pipe sizes.

Pipe size	Number of gallons
¾ inch	2.8
1 inch	4.5
1¼ inch	7.8
1½ inch	11.5
2 inch	18

FIGURE A.92 Fluid volume of pipe contents for polybutylene pipe (computed on the number of gallons per 100 feet of pipe).

Pipe size	Number of gallons
1 inch	4.1
1¼ inch	6.4
1½ inch	9.2

FIGURE A.94 Fluid volume of pipe contents for copper pipe (computed on the number of gallons per 100 feet of pipe).

Material	Weight in pounds per cubic inch	Weight in pounds per cubic foot
Aluminum	.093	160
Antimony	.2422	418
Brass	.303	524
Bronze	.320	552
Chromium	.2348	406
Copper	.323	558
Gold	.6975	1205
Iron (cast)	.260	450
Iron (wrought)	.2834	490
Lead	.4105	710
Manganese	.2679	463
Mercury	.491	849
Molybdenum	.309	534
Monel	.318	550
Platinum	.818	1413
Steel (mild)	.2816	490
Steel (stainless)	.277	484
Tin	.265	459
Titanium	.1278	221
Zinc	.258	446

FIGURE A.93 Weights of various materials.

GPM	Liters/Minute
1	3.75
2	6.50
3	11.25
4	15.00
5	18.75
6	22.50
7	26.25
8	30.00
9	33.75
10	37.50

FIGURE A.95 Flow rate conversion from gallons per minute (GPM) to approximate liters per minute.

Inches	Decimal of an inch
1/64	0.015625
1/32	0.03125
3/64	0.046875
1/16	0.0625
5/64	0.078125
3/32	0.09375
7/64	0.109375
1/8	0.125
9/64	0.140625
5/32	0.15625
11/64	0.171875
3/16	0.1875
12/64	0.203125
7/32	0.21875
15/64	0.234375
1/4	0.25
17/64	0.265625
9/32	0.28125
19/64	0.296875
5/16	0.3125

Note: To find the decimal equivalent of a fraction, divide the numerator by the denominator.

FIGURE A.96 Decimal equivalents of fractions of an inch.

Inches	Decimal of a foot
1/8	0.01042
1/4	0.02083
3/8	0.03125
1/2	0.04167
5/8	0.05208
3/4	0.06250
7/8	0.07291
1	0.08333
1 1/8	0.09375
1 1/4	0.10417
1 3/8	0.11458
1 1/2	0.12500
1 5/8	0.13542
1 3/4	0.14583
1 7/8	0.15625
2	0.16666
2 1/8	0.17708
2 1/4	0.18750
2 3/8	0.19792
2 1/2	0.20833
2 5/8	0.21875
2 3/4	0.22917
2 7/8	0.23959
3	0.25000

Note: To change inches to decimals of a foot, divide by 12. To change decimals of a foot to inches, multiply by 12.

FIGURE A.97 Inches converted to decimals of feet.

Vacuum in inches of mercury	Boiling point
29	76.62
28	99.93
27	114.22
26	124.77
25	133.22
24	140.31
23	146.45
22	151.87
21	156.75
20	161.19
19	165.24
18	169.00
17	172.51
16	175.80
15	178.91
14	181.82
13	184.61
12	187.21
11	189.75
10	192.19
9	194.50
8	196.73
7	198.87
6	200.96
7	198.87
6	200.96
5	202.25
4	204.85
3	206.70
2	208.50
1	210.25

FIGURE A.98 Boiling points of water at various pressures.

Barometer (in. Hg)	Pressure (lb/sq in.)
28.00	13.75
28.25	13.88
28.50	14.00
28.75	14.12
29.00	14.24
29.25	14.37
29.50	14.49
29.75	14.61
29.921	14.696
30.00	14.74
30.25	14.86
30.50	14.98
30.75	15.10
31.00	15.23

Rule: Barometer in in. Hg × 0.49116 = lb/sq in.

FIGURE A.99 Atmospheric pressure per square inch.

To change	to	Multiply by
Inches	Feet	0.0833
Inches	Millimeters	25.4
Feet	Inches	12
Feet	Yards	0.3333
Yards	Feet	3
Square inches	Square feet	0.00694
Square feet	Square inches	144
Square feet	Square yards	0.11111
Square yards	Square feet	9
Cubic inches	Cubic feet	0.00058
Cubic feet	Cubic inches	1728
Cubic feet	Cubic yards	0.03703
Gallons	Cubic inches	231
Gallons	Cubic feet	0.1337
Gallons	Pounds of water	8.33
Pounds of water	Gallons	0.12004
Ounces	Pounds	0.0625
Pounds	Ounces	16
Inches of water	Pounds per square inch	0.0361
Inches of water	Inches of mercury	0.0735
Inches of water	Ounces per square inch	0.578
Inches of water	Pounds per square foot	5.2
Inches of mercury	Inches of water	13.6
Inches of mercury	Feet of water	1.1333
Inches of mercury	Pounds per square inch	0.4914
Ounces per square inch	Inches of mercury	0.127
Ounces per square inch	Inches of water	1.733
Pounds per square inch	Inches of water	27.72
Pounds per square inch	Feet of water	2.310
Pounds per square inch	Inches of mercury	2.04
Pounds per square inch	Atmospheres	0.0681
Feet of water	Pounds per square inch	0.434
Feet of water	Pounds per square foot	62.5
Feet of water	Inches of mercury	0.8824
Atmospheres	Pounds per square inch	14.696
Atmospheres	Inches of mercury	29.92
Atmospheres	Feet of water	34
Long tons	Pounds	2240
Short tons	Pounds	2000
Short tons	Long tons	0.89295

FIGURE A.100 Measurement conversion factors.

U.S.	Metric
0.001 in.	0.025 mm
1 in.	25.400 mm
1 ft	30.48 cm
1 ft	0.3048 m
1 yd	0.9144 m
1 mi	1.609 km
1 in.^{2}	6.4516 cm^2
1 ft^2	0.0929 m^2
1 yd^2	0.8361 m^2
1 a	0.4047 ha
1 mi^2	2.590 km^2
1 in.^{3}	16.387 cm^3
1 ft^3	0.0283 m^3
1 yd^3	0.7647 m^3
1 U.S. oz	29.57 ml
1 U.S. p	0.4732 l
1 U.S. gal	3.785 l
1 oz	28.35 g
1 lb	0.4536 kg

FIGURE A.101 Metric conversion table.

Grade	Ratio	Material needed for each cubic yard of concrete
Strong—watertight, exposed to weather and moderate wear	1:2¼:3	6 bags cement 14 ft^3 sand (0.52 yd^3) 18 ft^3 stone (0.67 yd^3)
Moderate—moderate strength, not exposed	1:2¾:4	5 bags cement 14 ft^3 sand (0.52 yd^3) 20 ft^3 stone (0.74 yd^3)
Economy—massive areas, low strength	1:3:5	4½ bags cement 13 ft^3 sand (0.48 yd^3) 22 ft^3 stone (0.82 yd^3)

FIGURE A.102 Formulas for concrete.

U.S.	Metric
2/12	50/300
4/12	100/300
6/12	150/300
8/12	200/300
10/12	250/300
12/12	300/300

FIGURE A.103 Roof pitches.

1 ft^3	approx. 100 lb
1 yd^3	2700 lb
1 t	¾ yd or 20 ft^3
Average shovelful	15 lb
12-qt pail	40 lb

FIGURE A.104 Volume-to-weight conversions for sand.

Minutes	Decimal of a degree	Minutes	Decimal of a degree
1	0.0166	16	0.2666
2	0.0333	17	0.2833
3	0.0500	18	0.3000
4	0.0666	19	0.3166
5	0.0833	20	0.3333
6	0.1000	21	0.3500
7	0.1166	22	0.3666
8	0.1333	23	0.3833
9	0.1500	24	0.4000
10	0.1666	25	0.4166
11	0.1833		
12	0.2000		
13	0.2166		
14	0.2333		
15	0.2500		

FIGURE A.105 Minutes converted to decimal of a degree.

°C	Base temperature	°F
	−100°–30°	
−73	−100	−148
−68	−90	−130
−62	−80	−112
−57	−70	−94
−51	−60	−76
−46	−50	−58
−40	−40	−40
−34.4	−30	−22
−28.9	−20	−4
−23.3	−10	14
−17.8	0	32
−17.2	1	33.8
−16.7	2	35.6
−16.1	3	37.4
−15.6	4	39.2
−15.0	5	41.0
−14.4	6	42.8
−13.9	7	44.6
−13.3	8	46.4
−12.8	9	48.2
−12.2	10	50.0
−11.7	11	51.8
−11.1	12	53.6
−10.6	13	55.4
−10.0	14	57.2

°C	Base temperature	°F
	31°–71°	
−0.6	31	87.8
0	32	89.6
0.6	33	91.4
1.1	34	93.2
1.7	35	95.0
2.2	36	96.8
2.8	37	98.6
3.3	38	100.4
3.9	39	102.2
4.4	40	104.0
5.0	41	105.8
5.6	42	107.6
6.1	43	109.4
6.7	44	111.2
7.2	45	113.0

FIGURE A.106 Temperature conversion.

°C	Base temperature	°F
	31°–71°	
7.8	46	114.8
8.3	47	116.6
8.9	48	118.4
9.4	49	120.0
10.0	50	122.0
10.6	51	123.8
11.1	52	125.6
11.7	53	127.4
12.2	54	129.2
12.8	55	131.0

°C	Base temperature	°F
	72°–212°	
22.2	72	161.6
22.8	73	163.4
23.3	74	165.2
23.9	75	167.0
24.4	76	168.8
25.0	77	170.6
25.6	78	172.4
26.1	79	174.2
26.7	80	176.0
27.8	81	177.8
28.3	82	179.6
28.9	83	181.4
29.4	84	183.2
30.0	85	185.0
30.6	86	186.8
31.1	87	188.6
31.7	88	190.4
32.2	89	192.2
32.8	90	194.0
33.3	91	195.8
33.9	92	197.6
34.4	93	199.4
35.0	94	201.2
35.6	95	203.0
36.1	96	204.8

FIGURE A.106 Temperature conversion *(continued)*.

°C	Base temperature	°F
	213°–620°	
104	220	248
110	230	446
116	240	464
121	250	482
127	260	500
132	270	518
138	280	536
143	290	554
149	300	572
154	310	590
160	320	608
166	330	626
171	340	644
177	350	662
182	360	680

FIGURE A.106 Temperature conversion *(continued)*.

°C	Base temperature	°F
	213°–620°	
188	370	698
193	380	716
199	390	734
204	400	752
210	410	770
216	420	788
221	430	806
227	440	824
232	450	842
238	460	860

°C	Base temperature	°F
	621°–1000°	
332	630	1166
338	640	1184
343	650	1202
349	660	1220
354	670	1238
360	680	1256
366	690	1274
371	700	1292
377	710	1310
382	720	1328
388	730	1346
393	740	1364
399	750	1382
404	760	1400
410	770	1418
416	780	1436
421	790	1454
427	800	1472
432	810	1490
438	820	1508
443	830	1526
449	840	1544
454	850	1562
460	860	1580
466	870	1598

FIGURE A.106 Temperature conversion *(continued)*.

To change	to	Multiply by
Inches	Feet	0.0833
Inches	Millimeters	25.4
Feet	Inches	12
Feet	Yards	0.3333
Yards	Feet	3
Square inches	Square feet	0.00694
Square feet	Square inches	144
Square feet	Square yards	0.11111
Square yards	Square feet	9
Cubic inches	Cubic feet	0.00058
Cubic feet	Cubic inches	1728
Cubic feet	Cubic yards	0.03703
Cubic yards	Cubic feet	27
Cubic inches	Gallons	0.00433
Cubic feet	Gallons	7.48
Gallons	Cubic inches	231
Gallons	Cubic feet	0.1337
Gallons	Pounds of water	8.33
Pounds of water	Gallons	0.12004
Ounces	Pounds	0.0625
Pounds	Ounces	16
Inches of water	Pounds per square inch	0.0361
Inches of water	Inches of mercury	0.0735
Inches of water	Ounces per square inch	0.578
Inches of water	Pounds per square foot	5.2
Inches of mercury	Inches of water	13.6
Inches of mercury	Feet of water	1.1333
Inches of mercury	Feet of water	0.4914
Ounces per square inch	Pounds per square inch	0.127
Ounces per square inch	Inches of mercury	1.733
Pounds per square inch	Inches of water	27.72
Pounds per square inch	Feet of water	2.310
Pounds per square inch	Inches of mercury	2.04
Pounds per square inch	Atmospheres	0.0681
Feet of water	Pounds per square inch	0.434
Feet of water	Pounds per square foot	62.5
Feet of water	Inches of mercury	0.8824
Atmospheres	Pounds per square inch	14.696
Atmospheres	Inches of mercury	29.92
Atmospheres	Feet of water	34
Long tons	Pounds	2240
Short tons	Pounds	2000
Short tons	Long tons	0.89295

FIGURE A.107 Useful multipliers.

Ounces	Kilograms
1	0.028
2	0.057
3	0.085
4	0.113
5	0.142
6	0.170
7	0.198
8	0.227
9	0.255
10	0.283
11	0.312
12	0.340
13	0.369
14	0.397
15	0.425
16	0.454

FIGURE A.108 Ounces to kilograms.

Pounds	Kilograms
1	0.454
2	0.907
3	1.361
4	1.814
5	2.268
6	2.722
7	3.175
8	3.629
9	4.082
10	4.536
25	11.34
50	22.68
75	34.02
100	45.36

FIGURE A.109 Pounds to kilograms.

Gallons per minute	Liters per minute
1	3.75
2	6.50
3	11.25
4	15.00
5	18.75
6	22.50
7	26.25
8	30.00
9	33.75
10	37.50

FIGURE A.110 Flow-rate conversion.

1 GMP 0.134 cu ft/min
1 cu ft/min (cfm) 448.8 gal/hr (gph)

Feet per second	Meters per second
1	0.3050
2	0.610
3	0.915
4	1.220
5	1.525
6	1.830
7	2.135
8	2.440
9	2.754
10	3.050

FIGURE A.111 Flow-rate equivalents.

Pounds per square foot	Kilopascals
1	0.0479
2	0.0958
3	0.1437
4	0.1916
5	0.2395
6	0.2874
7	0.3353
8	0.3832
9	0.4311
10	0.4788
25	1.1971
50	2.394
75	3.5911
100	4.7880

FIGURE A.112 Pounds per square foot to kilopascals.

Pounds per square inch	Kilopascals
1	6.895
2	13.790
3	20.685
4	27.580
5	34.475
6	41.370
7	48.265
8	55.160
9	62.055
10	68.950
25	172.375
50	344.750
75	517.125
100	689.500

FIGURE A.113 Pounds per square inch to kilopascals.

Feet head	Pounds per square inch	Feet head	Pounds per square inch
1	0.43	50	21.65
2	0.87	60	25.99
3	1.30	70	30.32
4	1.73	80	34.65
5	2.17	90	38.98
6	2.60	100	43.34
7	3.03	110	47.64
8	3.46	120	51.97
9	3.90	130	56.30
10	4.33	140	60.63
15	6.50	150	64.96
20	8.66	160	69.29
25	10.83	170	73.63
30	12.99	180	77.96
40	17.32	200	86.62

FIGURE A.114 Water feet head to pounds per square inch.

1/32	0.03125
1/16	0.0625
3/32	0.09375
1/8	0.125
5/32	0.15625
3/16	0.1875
7/32	0.21875
1/4	0.25
9/32	0.28125
5/16	0.3125
11/32	0.34375
3/8	0.375
13/32	0.40625
7/16	0.4375
15/32	0.46875
1/2	0.500
17/32	0.53125
9/16	0.5625
19/32	0.59375
5/8	0.625
21/32	0.65625
11/16	0.6875
23/32	0.71875
3/4	0.75
25/32	0.78125
13/16	0.8125
27/32	0.84375
7/8	0.875
29/32	0.90625
15/16	0.9375
31/32	0.96875
1	1.000

FIGURE A.115 Decimal equivalents of an inch.

Pounds per square inch	Feet head
1	2.31
2	4.62
3	6.93
4	9.24
5	11.54
6	13.85
7	16.16
8	18.47
9	20.78
10	23.09
15	34.63
20	46.18
25	57.72
30	69.27
40	92.36
50	115.45
60	138.54
70	161.63
80	184.72
90	207.81
100	230.90
110	253.98
120	277.07
130	300.16
140	323.25
150	346.34
160	369.43
170	392.52
180	415.61
200	461.78
250	577.24
300	692.69
350	808.13
400	922.58
500	1154.48
600	1385.39
700	1616.30
800	1847.20
900	2078.10
1000	2309.00

FIGURE A.116 Water pressure in pounds with equivalent feet heads.

Outside design temperature	= average of lowest recorded temperature in each month from October to March
Inside design temperature	= 70°F or as specified by owner

A degree day is one day multiplied by the number of Fahrenheit degrees the mean temperature is below 65°F. The number of degree days in a year is a good guideline for designing heating and insulation systems.

FIGURE A.117 Design temperature.

To find	Multiply	By
Microns	Mils	25.4
Centimeters	Inches	2.54
Meters	Feet	0.3048
Meters	Yards	0.19144
Kilometers	Miles	1.609344
Grams	Ounces	28.349523
Kilograms	Pounds	0.4539237
Liters	Gallons (U.S.)	3.7854118
Liters	Gallons (imperial)	4.546090
Milliliters (cc)	Fluid ounces	29.573530
Milliliters (cc)	Cubic inches	16.387064
Square centimeters	Square inches	6.4516
Square meters	Square feet	0.09290304
Square meters	Square yards	0.83612736
Cubic meters	Cubic feet	2.8316847×10^{-2}
Cubic meters	Cubic yards	0.76455486
Joules	Btu	1054.3504
Joules	Foot-pounds	1.35582
Kilowatts	Btu per minute	0.01757251
Kilowatts	Foot-pounds per minute	2.2597×10^{-5}
Kilowatts	Horsepower	0.7457
Radians	Degrees	0.017453293
Watts	Btu per minute	17,5725

FIGURE A.118 Factors used in converting from customary (U.S.) units to metric units.

Horsepower	Kilowatt
1/20	0.025
1/16	0.05
1/8	0.1
1/6	0.14
1/4	0.2
1/3	0.28
1/2	0.4
1	0.8
1½	1.1

FIGURE A.119 Metric motor ratings.

1 ft^3	62.4 lbs
1 ft^3	7.48 gal
1 gal	8.33 lbs
1 gal	0.1337 ft^3

FIGURE A.120 Water volume to weight.

1 ft^3 at 50°F weighs 62.41 lb
1 gal at 50°F weighs 8.34 lb
1 ft^3 of ice weighs 57.2 lb
Water is at its greatest density at 39.2°F
1 ft^3 at 39.2°F weighs 62.43 lb

FIGURE A.121 Water weight.

Unit	Symbol
Length	
Millimeter	mm
Centimeter	cm
Meter	m
Kilometer	km
Area	
Square millimeter	mm^2
Square centimeter	cm^2
Square decimeter	dm^2
Square meter	m^2
Square kilometer	km^2
Volume	
Cubic centimeter	cm^3
Cubic decimeter	dm^3
Cubic meter	m^3
Mass	
Milligram	mg
Gram	g
Kilogram	kg
Tonne	t
Temperature	
Degrees Celsius	°C
Kelvin	K
Time	
Second	s
Plane angle	
Radius	rad
Force	
Newton	N
Energy, work, quantity of heat	
Joule	J
Kilojoule	kJ
Megajoule	MJ
Power, heat flow rate	
Watt	W
Kilowatt	kW
Pressure	
Pascal	Pa
Kilopascal	kPa
Megapascal	MPa
Velocity, speed	
Meter per second	m/s
Kilometer per hour	km/h

FIGURE A.122 Metric abbreviations.

Watt output	Millihorsepower (MHP)	Fractional HP
0.746	1	1/1000
1.492	2	1/500
2.94	4	1/250
4.48	6	1/170
5.97	8	1/125
7.46	10	1/100
9.33	12.5	1/80
10.68	14.3	1/70
11.19	15	1/65
11.94	16	1/60
14.92	20	1/50
18.65	25	1/40
22.38	30	1/35
24.90	33	1/30

FIGURE A.124 Motor power output comparison.

Length	
1 in.	25.4 mm
1 ft	0.3048 m
1 yd	0.9144 m
1 mi	1.609 km
Mass	
1 lb	0.454 kb
1 U.S. short ton	0.9072 t
Area	
1 ft^2	0.092 m^2
1 yd^2	0.836 m^2
1 a	0.404 ha
Capacity, liquid	
1 ft^3	0.028 m^3
Capacity, dry	
1 yd^3	0.764 m^3
Volume, liquid	
1 qt	0.946 l
1 gal	3.785 l
Heat	
1 Btu	1055 joule (J)
1 Btu/hr	0.293 watt (W)

FIGURE A.125 Measurements conversions (imperial to metric).

Inch scale		Metric scale
¼	is closest to	1:50
⅛	is closest to	1:100

FIGURE A.123 Scales used for building plans.

Inch scale		Metric scale
1/16	is closest to	1:200

FIGURE A.126 Scale used for site plans.

16 oz = 1 lb
100 lb = 1 cwt
20 cwt = 1 ton
1 ton = 2000 lb
1 long ton = 2240 lb
16 dr = 1 oz
16 oz = 1 lb
100 lb = 1 cwt
20 cwt = 1 ton = 2000 lb

FIGURE A.127 Avoirdupois weight.

10 milligrams (mg)	1 centigram
10 centigrams (cg)	1 decigram
10 decigrams (dg)	1 gram
10 grams (g)	1 dekagram
10 dekagrams (dkg)	1 hectogram
10 hectograms (hg)	1 kilogram
10 kilograms (kg)	1 myriagram
10 myriagrams (myg)	1 quintal

FIGURE A.128 Metric weight measure.

60 s	1 min
60 min	1°
360°	1 circle

FIGURE A.129 Circular measure.

1 gal	0.133681 ft^3
1 gal	231 $in.^3$

FIGURE A.130 Volume measure equivalents.

Temperature (°F)	Steel	Cast iron	Brass and copper
0	0	0	0
20	0.15	0.10	0.25
40	0.30	0.25	0.45
60	0.45	0.40	0.65
80	0.60	0.55	0.90
100	0.75	0.70	1.15
120	0.90	0.85	1.40
140	1.10	1.00	1.65
160	1.25	1.15	1.90
180	1.45	1.30	2.15
200	1.60	1.50	2.40
220	1.80	1.65	2.65
240	2.00	1.80	2.90
260	2.15	1.95	3.15
280	2.35	2.15	3.45
300	2.50	2.35	3.75
320	2.70	2.50	4.05
340	2.90	2.70	4.35
360	3.05	2.90	4.65
380	3.25	3.10	4.95
400	3.45	3.30	5.25
420	3.70	3.50	5.60
440	3.95	3.75	5.95
460	4.20	4.00	6.30
480	4.45	4.25	6.65
500	4.70	4.45	7.05

FIGURE A.131 Steam pipe expansion (inches increase per 100 inches).

Single-family dwellings; number of bedrooms	Multiple dwelling units or apartments; one bedroom each	Other uses; maximum fixture-units served	Minimum septic tank capacity in gallons
1–3		20	1000
4	2	25	1200
5–6	3	33	1500
7–8	4	45	2000
	5	55	2250
	6	60	2500
	7	70	2750
	8	80	3000
	9	90	3250
	10	100	3500

FIGURE A.132 Common septic tank capacities.

Appliance	Size (inches)
Clothes washer	2
Bathtub with or without shower	1½
Bidet	1½
Dental unit or cuspidor	1¼
Drinking fountain	1¼
Dishwasher, domestic	1½
Dishwasher, commercial	2
Floor drain	2, 3, or 4
Lavatory	1¼
Laundry tray	1½
Shower stall, domestic	2
Sinks:	
Combination, sink and tray (with disposal unit)	1½
Combination, sink and tray (with one trap)	1½
Domestic, with or without disposal unit	1½
Surgeon's	1½
Laboratory	1½
Flushrim or bedpan washer	3
Service sink	2 or 3
Pot or scullery sink	2
Soda fountain	1½
Commercial, flat rim, bar, or counter sink	1½
Wash sinks circular or multiple	1½
Urinals:	
Pedestal	3
Wall-hung	1½ or 2
Trough (per 6-ft section)	1½
Stall	2
Water closet	3

FIGURE A.133 Common trap sizes.

Category	Estimated water usage per day
Barber shop	100 gal per chair
Beauty shop	125 gal per chair
Boarding school, elementary	75 gal per student
Boarding school, secondary	100 gal per student
Clubs, civic	3 gal per person
Clubs, country	25 gal per person
College, day students	25 gal per student
College, junior	100 gal per student
College, senior	100 gal per student
Dentist's office	750 gal per chair
Department store	40 gal per employee
Drugstore	500 gal per store
Drugstore with fountain	2000 gal per store
Elementary school	16 gal per student
Hospital	400 gal per patient
Industrial plant	30 gal per employee + process water
Junior and senior high school	25 gal per student
Laundry	2000–20,000 gal
Launderette	1000 gal per unit
Meat market	5 gal per 100 ft^2 of floor area
Motel or hotel	125 gal per room
Nursing home	150 gal per patient
Office building	25 gal per employee
Physician's office	200 gal per examining room
Prison	60 gal per inmate
Restaurant	20–120 gal per seat
Rooming house	100 gal per tenant
Service station	600–1500 gal per stall
Summer camp	60 gal per person
Theater	3 gal per seat

FIGURE A.134 Estimating guidelines for daily water usage.

Unit	Pounds per square inch	Feet of water	Meters of water	Inches of mercury	Atmospheres
1 pound per square inch	1.0	2.31	0.704	2.04	0.0681
1 foot of water	0.433	1.0	0.305	0.882	0.02947
1 meter of water	1.421	3.28	1.00	2.89	0.0967
1 inch of mercury	0.491	1.134	0.3456	1.00	0.0334
1 atmosphere (sea level)	14.70	33.93	10.34	29.92	1.0000

FIGURE A.135 Conversion of water values.

To convert	Multiply by	To obtain
	A	
acres	4.35×10^4	sq. ft.
acres	4.047×10^3	sq. meters
acre-feet	4.356×10^4	cu. feet
acre-feet	3.259×10^5	gallons
atmospheres	2.992×10^1	in. of mercury (at 0°C.)
atmospheres	1.0333	kgs./sq. cm.
atmospheres	1.0333×10^4	kgs./sq. meter
atmospheres	1.47×10^1	pounds/sq. in.
	B	
barrels (U.S., liquid)	3.15×10^1	gallons
barrels (oil)	4.2×10^1	gallons (oil)
bars	9.869×10^{-1}	atmospheres
btu	7.7816×10^2	foot-pounds
btu	3.927×10^{-4}	horsepower-hours
btu	2.52×10^{-1}	kilogram-calories
btu	2.928×10^{-4}	kilowatt-hours
btu/hr.	2.162×10^{-1}	ft. pounds/sec.
btu/hr.	3.929×10^{-4}	horsepower
btu/hr.	2.931×10^{-1}	watts
btu/min.	1.296×10^1	ft.-pounds/sec.
btu/min.	1.757×10^{-2}	kilowatts
	C	
centigrade (degrees)	(°C × 9/5) + 32	fahrenheit (degrees)
centigrade (degrees)	°C + 273.18	kelvin (degrees)
centigrams	$1. \times 10^{-2}$	grams
centimeters	3.281×10^{-2}	feet
centimeters	3.937×10^{-1}	inches
centimeters	$1. \times 10^{-5}$	kilometers
centimeters	$1. \times 10^{-2}$	meters
centimeters	$1. \times 10^1$	millimeters
centimeters	3.937×10^2	mils
centimeters of mercury	1.316×10^{-2}	atmospheres
centimeters of mercury	4.461×10^{-1}	ft. of water
centimeters of mercury	1.934×10^{-1}	pounds/sq. in.
centimeters/sec.	1.969	feet/min.
centimeters/sec.	3.281×10^{-2}	feet/sec.
centimeters/sec.	6.0×10^{-1}	meters/min.
centimeters/sec./sec.	3.281×10^{-2}	ft./sec./sec.
cubic centimeters	3.531×10^{-5}	cubic ft.
cubic centimeters	6.102×10^{-2}	cubic in.
cubic centimeters	1.0×10^{-6}	cubic meters
cubic centimeters	2.642×10^{-4}	gallons (U.S. liquid)
cubic centimeters	2.113×10^{-3}	pints (U.S. liquid)
cubic centimeters	1.057×10^{-3}	quarts (U.S. liquid)

FIGURE A.136 Measurement conversions.

To convert	Multiply by	To obtain
cubic feet	2.8320×10^4	cu. cms.
cubic feet	1.728×10^3	cu. inches
cubic feet	2.832×10^{-2}	cu. meters
cubic feet	7.48052	gallons (U.S. liquid)
cubic feet	5.984×10^1	pints (U.S. liquid)
cubic feet	2.992×10^1	quarts (U.S. liquid)
cubic feet/min.	4.72×10^1	cu. cms./sec.
cubic feet/min.	1.247×10^{-1}	gallons/sec.
cubic feet/min.	4.720×10^{-1}	liters/sec.
cubic feet/min.	6.243×10^1	pounds water/min.
cubic feet/sec.	6.46317×10^{-1}	million gals./day
cubic feet/sec.	4.48831×10^2	gallons/min.
cubic inches	5.787×10^{-4}	cu. ft.
cubic inches	1.639×10^{-5}	cu. meters
cubic inches	2.143×10^{-5}	cu. yards
cubic inches	4.329×10^{-3}	gallons
	D	
degrees (angle)	1.745×10^{-2}	radians
degrees (angle)	3.6×10^3	seconds
degrees/sec.	2.778×10^{-3}	revolutions/sec.
dynes/sq. cm.	4.015×10^{-4}	in. of water (at 4°C.)
dynes	1.020×10^{-6}	kilograms
dynes	2.248×10^{-6}	pounds
	F	
fathoms	1.8288	meters
fathoms	6.0	feet
feet	3.048×10^1	centimeters
feet	3.048×10^{-1}	meters
feet of water	2.95×10^{-2}	atmospheres
feet of water	3.048×10^{-2}	kgs./sq. cm.
feet of water	6.243×10^1	pounds/sq. ft.
feet/min.	5.080×10^{-1}	cms./sec.
feet/min.	1.667×10^{-2}	feet/sec.
feet/min.	3.048×10^{-1}	meters/min.
feet/min.	1.136×10^{-2}	miles/hr.
feet/sec.	1.829×10^1	meters/min.
feet/100 feet	1.0	per cent grade
foot-pounds	1.286×10^{-3}	btu
foot-pounds	1.356×10^7	ergs
foot-pounds	3.766×10^{-7}	kilowatt-hrs.
foot-pounds/min.	1.286×10^{-3}	btu/min.
foot-pounds/min.	3.030×10^{-5}	horsepower
foot-pounds/min.	3.241×10^{-4}	kg.-calories/min.
foot-pounds/sec.	4.6263	btu/hr.
foot-pounds/sec.	7.717×10^{-2}	btu/min.
foot-pounds/sec.	1.818×10^{-3}	horsepower
foot-pounds/sec.	1.356×10^{-3}	kilowatts
furlongs	1.25×10^{-1}	miles (U.S.)

FIGURE A.136 Measurement conversions *(continued)*.

To convert	Multiply by	To obtain
	G	
gallons	3.785×10^3	cu. cms.
gallons	1.337×10^{-1}	cu. feet
gallons	2.31×10^2	cu. inches
gallons	3.785×10^{-3}	cu. meters
gallons	4.951×10^{-3}	cu. yards
gallons	3.785	liters
gallons (liq. br. imp.)	1.20095	gallons (U.S. liquid)
gallons (U.S.)	$8,3267 \times 10^{-1}$	gallons (imp.)
gallons of water	8.337	pounds of water
gallons/min.	2.228×10^{-3}	cu. feet/sec.
gallons/min.	6.308×10^{-2}	liters/sec.
gallons/min.	8.0208	cu. feet/hr.
	H	
horsepower	4.244×10^1	btu/min.
horsepower	3.3×10^4	foot-lbs./min.
horsepower	5.50×10^2	foot-lbs./sec.
horsepower (metric)	9.863×10^{-1}	horsepower
horsepower	1.014	horsepower (metric)
horsepower	7.457×10^{-1}	kilowatts
horsepower	7.457×10^2	watts
horsepower (boiler)	3.352×10^4	btu/hr.
horsepower (boiler)	9.803	kilowatts
horsepower-hours	2.547×10^3	btu
horsepower-hours	1.98×10^6	foot-lbs.
horsepower-hours	6.4119×10^5	gram-calories
hours	5.952×10^{-3}	weeks
	I	
inches	2.540	centimeters
inches	2.540×10^{-2}	meters
inches	1.578×10^{-5}	miles
inches	2.54×10^1	millimeters
inches	1.0×10^3	mils
inches	2.778×10^{-2}	yards
inches of mercury	3.342×10^{-2}	atmospheres
inches of mercury	1.133	feet of water
inches of mercury	3.453×10^{-2}	kgs./sq. cm.
inches of mercury	3.453×10^2	kgs./sq. meter
inches of mercury	7.073×10^1	pounds/sq. ft.
inches of mercury	4.912×10^{-1}	pounds/sq. in.
in. of water (at 4°C.)	7.355×10^{-2}	inches of mercury
in. of water (at 4°C.)	2.54×10^{-3}	kgs./sq. cm.
in. of water (at 4°C.)	5.204	pounds/sq. ft.
in. of water (at 4°C.)	3.613×10^{-2}	pounds/sq. in.

FIGURE A.136 Measurement conversions *(continued)*.

To convert	Multiply by	To obtain
	J	
joules	9.486×10^{-4}	btu
joules/cm.	1.0×10^{7}	dynes
joules/cm.	1.0×10^{2}	joules/meter (newtons)
joules/cm.	2.248×10^{1}	pounds
	K	
kilograms	9.80665×10^{5}	dynes
kilograms	1.0×10^{3}	grams
kilograms	2.2046	pounds
kilograms	9.842×10^{-4}	tons (long)
kilograms	1.102×10^{-3}	tons (short)
kilograms/sq. cm.	9.678×10^{-1}	atmospheres
kilograms/sq. cm.	3.281×10^{1}	feet of water
kilograms/sq. cm.	2.896×10^{1}	inches of mercury
kilograms/sq. cm.	1.422×10^{1}	pounds/sq. in.
kilometers	1.0×10^{5}	centimeters
kilometers	3.281×10^{3}	feet
kilometers	3.937×10^{4}	inches
kilometers	1.0×10^{3}	meters
kilometers	6.214×10^{-1}	miles (statute)
kilometers	5.396×10^{-1}	miles (nautical)
kilometers	1.0×10^{6}	millimeters
kilowatts	5.692×10^{1}	btu/min.
kilowatts	4.426×10^{4}	foot-lbs./min.
kilowatts	7.376×10^{2}	foot-lbs./sec.
kilowatts	1.341	horsepower
kilowatts	1.434×10^{1}	kg.-calories/min.
kilowatts	1.0×10^{3}	watts
kilowatt-hrs.	3.413×10^{3}	btu
kilowatt-hrs.	2.655×10^{6}	foot-lbs.
kilowatt-hrs.	8.5985×10^{3}	gram calories
kilowatt-hrs.	1.341	horsepower-hours
kilowatt-hrs.	3.6×10^{6}	joules
kilowatt-hrs.	8.605×10^{2}	kg.-calories
kilowatt-hrs.	8.5985×10^{3}	kg.-meters
kilowatt-hrs.	2.275×10^{1}	pounds of water raised from 62° to 212°F.
	L	
links (engineers)	1.2×10^{1}	inches
links (surveyors)	7.92	inches
liters	1.0×10^{3}	cu. cm.
liters	6.102×10^{1}	cu. inches
liters	1.0×10^{-3}	cu. meters
liters	2.642×10^{-1}	gallons (U.S. liquid)
liters	2.113	pints (U.S. liquid)
liters	1.057	quarts (U.S. liquid)

FIGURE A.136 Measurement conversions *(continued)*.

To convert	Multiply by	To obtain
	M	
meters	1.0×10^{2}	centimeters
meters	3.281	feet
meters	3.937×10^{1}	inches
meters	1.0×10^{-3}	kilometers
meters	5.396×10^{-4}	miles (nautical)
meters	6.214×10^{-4}	miles (statute)
meters	1.0×10^{3}	millimeters
meters/min.	1.667	cms./sec.
meters/min.	3.281	feet/min.
meters/min.	5.468×10^{-2}	feet/sec.
meters/min.	6.0×10^{-2}	kms./hr.
meters/min.	3.238×10^{-2}	knots
meters/min.	3.728×10^{-2}	miles/hr.
meters/sec.	1.968×10^{2}	feet/min.
meters/sec.	3.281	feet/sec.
meters/sec.	3.6	kilometers/hr.
meters/sec.	6.0×10^{-2}	kilometers/min.
meters/sec.	2.237	miles/hr.
meters/sec.	3.728×10^{-2}	miles/min.
miles (nautical)	6.076×10^{3}	feet
miles (statute)	5.280×10^{3}	feet
miles/hr.	8.8×10^{1}	ft./min.
millimeters	1.0×10^{-1}	centimeters
millimeters	3.281×10^{-3}	feet
millimeters	3.937×10^{-2}	inches
millimeters	1.0×10^{-1}	meters
minutes (time)	9.9206×10^{-5}	weeks
	O	
ounces	2.8349×10^{1}	grams
ounces	6.25×10^{-2}	pounds
ounces (fluid)	1.805	cu. inches
ounces (fluid)	2.957×10^{-2}	liters
	P	
parts/million	5.84×10^{-2}	grains/u.s. gal.
parts/million	7.016×10^{-2}	grains/imp. gal.
parts/million	8.345	pounds/million gal.
pints (liquid)	4.732×10^{2}	cubic cms.
pints (liquid)	1.671×10^{-2}	cubic ft.
pints (liquid)	2.887×10^{1}	cubic inches
pints (liqui)	4.732×10^{-4}	cubic meters
pints (liquid)	1.25×10^{-1}	gallons
pints (liquid)	4.732×10^{-1}	liters
pints (liquid)	5.0×10^{-1}	quarts (liquid)

FIGURE A.136 Measurement conversions *(continued).*

To convert	Multiply by	To obtain
pounds	2.56×10^{2}	drams
pounds	4.448×10^{5}	dynes
pounds	7.0×10^{1}	grains
pounds	4.5359×10^{2}	grams
pounds	4.536×10^{-1}	kilograms
pounds	1.6×10^{1}	ounces
pounds	3.217×10^{1}	pounds
pounds	1.21528	pounds (troy)
pounds of water	1.602×10^{-2}	cu. ft.
pounds of water	2.768×10^{1}	cu. inches
pounds of water	1.198×10^{-1}	gallons
pounds of water/min.	2.670×10^{-4}	cu. ft./sec.
pound-feet	1.356×10^{7}	cm.-dynes
pound-feet	1.3825×10^{4}	cm.-grams
pound-feet	1.383×10^{-1}	meter-kgs.
pounds/cu. ft.	1.602×10^{-2}	grams/cu. cm.
pounds/cu. ft.	5.787×10^{-4}	pounds/cu. inches
pounds/sq. in.	6.804×10^{-2}	atmospheres
pounds/sq. in.	2.307	feet of water
pounds/sq. in.	2.036	inches of mercury
pounds/sq. in.	7.031×10^{2}	kgs./sq. meter
pounds/sq. in.	1.44×10^{2}	pounds/sq. ft.
	Q	
quarts (dry)	6.72×10^{1}	cu. inches
quarts (liquid)	9.464×10^{2}	cu. cms.
quarts (liquid)	3.342×10^{-2}	cu. ft.
quarts (liquid)	5.775×10^{1}	cu. inches
quarts (liquid)	2.5×10^{-1}	gallons
	R	
revolutions	3.60×10^{2}	degrees
revolutions	4.0	quadrants
rods (surveyors' meas.)	5.5	yards
rods	1.65×10^{1}	feet
rods	1.98×10^{2}	inches
rods	3.125×10^{-3}	miles

FIGURE A.136 Measurement conversions *(continued).*

To convert	Multiply by	To obtain
	S	
slugs	3.217×10^{1}	pounds
square centimeters	1.076×10^{-3}	sq. feet
square centimeters	1.550×10^{-1}	sq. inches
square centimeters	1.0×10^{-4}	sq. meters
square centimeters	1.0×10^{2}	sq. millimeters
square feet	2.296×10^{-5}	acres
square feet	9.29×10^{2}	sq. cms.
square feet	1.44×10^{2}	sq. inches
square feet	9.29×10^{-2}	sq. meters
square feet	3.587×10^{-3}	sq. miles
square inches	6.944×10^{-3}	sq. ft.
square inches	6.452×10^{2}	sq. millimeters
square miles	6.40×10^{2}	acres
square miles	2.788×10^{7}	sq. ft.
square yards	2.066×10^{-4}	acres
square yards	8.361×10^{3}	sq. cms.
square yards	9.0	sq. ft.
square yards	1.296×10^{3}	sq. inches
	T	
temperature (°C.) +273	1.0	absolute temperature (°K.)
temperature (°C.) +17.78	1.8	temperature (°F.)
temperature (°F.) +460	1.0	absolute temperature (°R.)
temperature (°F.) −32	$5/9$	temperature (°C.)
tons (long)	2.24×10^{2}	pounds
tons (long)	1.12	tons (short)
tons (metric)	2.205×10^{5}	pounds
tons (short)	2.0×10^{3}	pounds
	W	
watts	3.4129	btu/hr.
watts	5.688×10^{-2}	btu/min.
watts	4.427×10^{1}	ft.-lbs/min.
watts	7.378×10^{-1}	ft.-lbs./sec
watts	1.341×10^{-3}	horsepower
watts	1.36×10^{-3}	horsepower (metric)
watts	1.0×10^{-3}	kilowatts
watt-hours	3.413	btu
watt-hours	2.656×10^{3}	foot-lbs.
watt-hours	1.341×10^{-3}	horsepower-hours
watt (international)	1.000165	watt (absolute)
weeks	1.68×10^{2}	hours
weeks	1.008×10^{4}	minutes
weeks	6.048×10^{5}	seconds

Source: *Pump Handbook* by I. J. Karassik et al. Copyright 1976, McGraw-Hill, Inc.

FIGURE A.136 Measurement conversions *(continued)*.

Contaminant	Suggested maximum level, mg/L
Calcium	2 (0.1 meq/L)
Magnesium	4 (0.3 meq/L)
Sodium	70 (3 meq/L)
Potassium	8 (0.2 meq/L)
Fluoride	0.2
Chlorine	0.5
Chloramines	0.1
Nitrate (N)	2
Sulfate	100
Copper, barium, zinc	0.1 each
Arsenic, lead, silver	0.005 each
Chromium	0.014
Cadmium	0.001
Selenium	0.09
Aluminum	0.01
Mercury	0.0002
Bacteria	200 (cfu/mL)

Source: Association for the Advancement of Medical Instrumentation (AAMI) "Hemodialysis Systems Standard," March 1990. Adopted by American National Standards Institute (ANSI), 1992.

FIGURE A.137 AAMI/ANSI water quality standards.

Nom. pipe size, in	Relative humidity, %														
	20			50			70			80			90		
	THK*	HG†	ST‡	THK	HG	ST	THK	HG	ST	THK	HG	ST	THK	HG	ST
0.50				0.5	2	66	0.5	2	66	0.5	2	66	1.0	2	68
0.75				0.5	2	67	0.5	2	67	0.5	2	67	0.5	2	67
1.00				0.5	3	66	0.5	3	66	0.5	3	66	1.0	2	68
1.25				0.5	3	66	0.5	3	66	0.5	3	66	1.0	3	67
1.50				0.5	4	65	0.5	4	65	0.5	4	65	1.0	3	67
2.00				0.5	5	66	0.5	5	66	0.5	5	66	1.0	3	67
2.50				0.5	5	65	0.5	5	65	0.5	5	65	1.0	4	67
3.00				0.5	7	65	0.5	7	65	0.5	7	65	1.0	4	67
3.50	Condensation control not required for this condition			0.5	8	65	0.5	8	65	0.5	8	65	1.0	4	68
4.00				0.5	8	65	0.5	8	65	0.5	8	65	1.0	5	67
5.00				0.5	10	65	0.5	10	65	0.5	10	65	1.0	6	67
6.00				0.5	12	65	0.5	12	65	0.5	12	65	1.0	7	67
8.00				1.0	9	67	1.0	9	67	1.0	9	67	1.0	9	67
10.00				1.0	11	67	1.0	11	67	1.0	11	67	1.0	11	67
12.00				1.0	12	67	1.0	12	67	1.0	12	67	1.0	12	67

*THK—Insulation thickness, inches.
†HG—Heat gain/lineal foot (pipe) 28 ft (flat).
‡ST—Surface temperature.

FIGURE A.138 Insulation thickness to prevent condensation, 500°F and 700°F ambient temperature.

Nom. pipe size, in	Relative humidity, %														
	20			50			70			80			90		
	THK*	HG†	ST‡	THK	HG	ST	THK	HG	ST	THK	HG	ST	THK	HG	ST
0.50				0.5	4	64	0.5	4	64	0.5	4	64	1.5	2	68
0.75				0.5	4	64	0.5	4	64	0.5	4	64	1.5	3	67
1.00				0.5	6	63	0.5	6	63	1.0	4	66	1.5	3	67
1.25				0.5	6	63	0.5	6	63	1.0	5	65	1.5	3	67
1.50				0.5	8	62	0.5	8	62	1.0	5	66	1.5	4	67
2.00				0.5	8	63	0.5	8	63	1.0	6	66	1.5	4	67
2.50				0.5	10	63	0.5	10	63	1.0	6	66	1.5	5	67
3.00				0.5	12	62	0.5	12	62	1.0	8	65	1.5	6	67
3.50	Condensation			0.5	14	61	0.5	14	61	1.0	7	66	1.5	6	67
4.00	control not			0.5	15	62	0.5	15	62	1.0	9	65	1.5	7	67
5.00	required for this			0.5	16	63	0.5	16	63	1.0	11	65	2.0	7	67
6.00	condition			0.5	22	61	0.5	22	61	1.0	13	65	2.0	8	67
8.00				1.0	16	65	1.0	16	65	1.0	16	65	2.0	10	67
10.00				1.0	20	65	1.0	20	65	1.0	20	65	2.0	11	67
12.00				1.0	22	65	1.0	22	65	1.0	22	65	2.0	13	67

*THK—Insulation thickness, inches.
†HG—Heat gain/lineal foot (pipe) 28 ft (flat).
‡ST—Surface temperature.

FIGURE A.139 Insulation thickness to prevent condensation, 340°F service temperature and 700°F ambient temperature.

Type of soil	Required sq. ft. of leaching area/100 gal. (m^2/L)	Maximum absorption capacity gals./sq. ft. of leaching area for a 24 hr. period (L/m^2)
1. Coarse sand or gravel	20 (.005)	5 (203.7)
2. Fine sand	25 (.006)	4 (162.9)
3. Sandy loam or sandy clay	40 (.010)	2.5 (101.9)
4. Clay with considerable sand or gravel	90 (.022)	1.10 (44.8)
5. Clay with small amount of sand or gravel	120 (.029)	0.83 (33.8)

FIGURE A.140 Design criteria of five typical soils.

Anticipated well yield, gpm	Nominal pump bowl size, in	Optimum well casing size, in	Smallest well casing size, in
Less than 100	4	6 I.D.	5 I.D.
75 to 175	5	8 I.D.	6 I.D.
150 to 400	6	10 I.D.	8 I.D.
350 to 650	8	12 I.D.	10 I.D.
600 to 900	10	14 I.D.	12 I.D.
850 to 1300	12	16 I.D.	14 I.D.
1200 to 1800	14	20 I.D.	16 I.D.
1600 to 3000	16	24 I.D.	20 I.D.

FIGURE A.141 Recommended well diameters.

	100°F	200°F	300°F	400°F	500°F	600°F
Fiberglass	0.26	0.30	0.34			
Polyurethane	0.16	0.16	0.16			
Calcium silicate	0.33	0.37	0.41	0.46	0.57	0.60
Cellular glass	0.39	0.47	0.55	0.64	0.74	0.85

Note: These are representative values per inch thickness for one square foot of area. Exact values should be confirmed by the insulation manufacturer.

FIGURE A.142 Heat loss in Btus through common insulation materials.

Fixture	Pressure, psi
Basin faucet	8
Basin faucet, self-closing	12
Sink faucet, ⅜ in (0.95 cm)	10
Sink faucet, ½ in (1.3 cm)	5
Dishwasher	15–25
Bathtub faucet	5
Laundry tub cock, ¼ in (0.64 cm)	5
Shower	12
Water closet flush tank	15
Water closet flush valve	15–20
Urinal flush valve	15
Garden hose, 50 ft (15 m), and sill cock	30
Water closet, blowout type	25
Urinal, blowout type	25
Water closet, low-silhouette tank type	30–40
Water closet, pressure tank	20–30

FIGURE A.143 Minimum acceptable operating pressures for various fixtures.

Use	Minimum temperature, °F
Lavatory:	
Hand washing	110
Shaving	110
Showers and tubs	110
Commercial and institutional laundry	180
Residential dishwashing and laundry	140
Commercial spray-type dishwashing as required by National Sanitation Foundation:	
Single or multiple tank hood or rack type:	
Wash	140
Final rinse	180 to 195
Single-tank conveyor type:	
Wash	160
Final rinse	180 to 195
Single-tank rack or door type:	
Single-temperature wash and rinse	165
Chemical sanitizing glasswasher:	
Wash	140
Rinse	75

FIGURE A.144 Minimum hot water temperature for plumbing fixtures and equipment.

Length of lateral in feet (m), 3" or 4" (76.2–101.6 mm) diameter		Length of main line in feet (m)								6" (152.4 mm) diameter					
		25 7.6	50 15.2	75 22.9	100 30.5	125 38.1	150 45.7	175 53.3	200 61	225 68.6	250 76.2	275 83.8	300 91.4	400 121.9	500 152.4
25	7.6	30	24	34	44	54	64	74	84	94	103	113	123	163	168
50	15.2	30	29	39	48	58	68	78	88	98	108	118	128	166	167
75	22.9	30	33	43	53	63	73	83	92	102	112	122	132	164	165
100	30.5	30	37	47	57	67	77	87	97	107	117	127	136	162	163
125	38.1	32	42	52	62	72	81	91	101	111	121	131	141	160	162
150	45.7	36	46	56	66	76	86	96	106	116	125	135	145	159	161
175	53.3	41	51	61	70	80	90	100	110	120	130	140	150	157	159
200	61	45	55	65	75	85	95	105	114	124	134	144	153	156	158
225	68.6	50	59	69	79	89	99	109	119	129	139	149	151	154	157
250	76.2	54	64	74	84	94	103	113	123	133	143	149	150	153	156
275	83.8	58	68	78	88	98	108	118	128	138	146	147	149	152	155
300	91.4	63	73	83	92	102	112	122	132	142	145	146	147	151	154
350	106.7	72	81	91	101	111	121	131	140	141	143	144	145	149	152
400	121.9	80	90	100	110	120	130	136	138	139	141	142	143	147	150
450	137.2	89	99	109	119	129	132	134	136	138	139	141	142	145	149
500	152.4	98	108	118	126	129	131	133	135	136	138	139	140	144	147

No holding time less than 30 seconds

Length of lateral in feet (m), 3" or 4" (76.2–101.6 mm) diameter		Length of main line in feet (m)								8" (203.2 mm) diameter					
		25 7.6	50 15.2	75 22.9	100 30.5	125 38.1	150 45.7	175 53.3	200 61	225 68.6	250 76.2	275 83.8	300 91.4	400 121.9	500 152.4
25	7.6	30	40	57	75	92	110	128	145	163	180	198	216	223	224
50	15.2	30	44	62	79	97	114	132	150	167	185	202	218	220	221
75	22.9	31	48	66	84	101	119	136	154	172	189	207	214	217	219
100	30.5	35	53	70	88	106	123	141	158	176	194	209	211	214	216
125	38.1	40	57	75	92	110	128	145	163	180	198	206	207	211	214
150	45.7	44	62	79	97	114	132	150	167	185	201	202	204	209	212
175	53.3	48	66	84	101	119	136	154	172	189	197	199	201	206	210
200	61	53	70	88	106	123	141	158	176	192	194	197	199	204	208
225	68.6	57	75	92	110	128	145	163	180	189	192	194	196	202	206
250	76.2	62	79	97	114	132	150	167	183	186	189	191	193	200	204
275	83.8	66	84	101	119	136	154	172	181	184	187	189	191	198	202
300	91.4	70	88	106	123	141	158	174	178	181	184	187	189	196	200
350	106.7	79	97	114	132	150	166	170	174	177	180	183	185	192	197
400	121.9	88	106	123	141	157	162	166	170	174	176	179	181	189	194
450	137.2	97	114	132	148	154	159	163	167	170	173	176	178	186	191
500	152.4	106	123	140	146	151	156	160	164	167	170	173	175	183	189

No holding time less than 30 seconds

FIGURE A.145 Low-pressure air test for building sewers.

Length of lateral in feet (m) 6" (152.4 mm) diameter		Length of main line in feet (m)								8" (203.2 mm) diameter					
		25	50	75	100	125	150	175	200	225	250	275	300	400	500
		7.6	15.2	22.9	30.5	38.1	45.7	53.3	61	68.6	76.2	83.8	91.4	121.9	152.4
25	7.6	30	45	63	80	98	116	133	151	168	186	204	221	224	225
50	15.2	37	55	73	90	108	126	143	161	178	196	214	220	222	223
75	22.9	47	65	83	100	118	135	153	171	188	206	217	217	220	221
100	30.5	57	75	93	110	128	145	163	181	198	214	214	215	218	220
125	38.1	67	85	102	120	138	155	173	190	208	211	212	213	216	218
150	45.7	77	95	112	130	148	165	182	200	207	209	210	211	214	217
175	53.3	87	105	122	140	157	175	192	204	206	207	208	209	214	215
200	61	97	114	132	150	167	185	201	202	204	205	206	207	211	214
225	68.6	107	124	142	160	177	195	199	201	203	204	205	206	210	213
250	76.2	117	134	152	169	187	195	198	199	201	202	203	204	209	212
275	83.8	127	144	162	179	192	194	196	198	200	201	202	204	208	210
300	91.4	136	154	172	187	190	192	195	196	198	200	201	202	207	209
350	106.7	156	174	181	185	187	190	193	194	196	198	199	200	205	208
400	121.9	173	178	181	184	186	189	191	192	194	196	197	198	203	206
450	137.2	173	177	180	183	185	187	189	190	192	194	195	196	201	204
500	152.4	173	177	180	182	184	186	188	189	191	192	193	194	200	203

No holding time less than 30 seconds

Length of lateral in feet (m) 4" (101.6 mm) diameter		Length of main line in feet (m)								6" (152.4 mm) diameter					
		25	50	75	100	125	150	175	200	225	250	275	300	400	500
		7.6	15.2	22.9	30.5	38.1	45.7	53.3	61	68.6	76.2	83.8	91.4	121.9	152.4
25	7.6	32	59	87	114	142	169	197	224	252	277	277	278	279	280
50	15.2	36	64	91	119	146	174	201	229	256	271	272	273	275	277
75	22.9	41	68	96	123	151	178	206	233	261	265	267	268	272	274
100	30.5	45	73	100	128	155	183	210	238	258	260	262	264	268	271
125	38.1	50	77	105	132	160	187	214	242	253	255	257	259	264	268
150	45.7	54	81	109	136	164	191	219	244	248	251	253	255	261	265
175	53.3	58	86	113	141	168	196	223	239	243	246	249	251	258	262
200	61	63	90	118	145	173	200	228	235	239	242	245	248	255	260
225	68.6	67	95	122	150	177	205	226	231	235	239	242	244	252	257
250	76.2	72	99	127	154	182	209	222	227	231	235	238	241	249	255
275	83.8	76	103	131	158	186	211	218	223	228	231	235	238	247	253
300	91.4	80	108	135	163	190	208	214	220	224	228	232	235	244	250
350	106.7	89	117	144	172	194	201	208	213	218	222	226	229	239	246
400	121.9	98	125	153	179	188	196	202	208	213	217	221	224	235	242
450	137.2	107	134	162	174	183	191	197	203	208	212	216	220	230	238
500	152.4	116	143	160	170	179	186	193	198	203	208	212	215	226	235

No holding time less than 30 seconds

FIGURE A.145 Low-pressure air test for building sewers *(continued)*.

Length of lateral in feet (m), 6" (152.4 mm) diameter — Length of main line in feet (m), 10" (254 mm) diameter

Lateral ft	Lateral m	25 7.6	50 15.2	75 22.9	100 30.5	125 38.1	150 45.7	175 53.3	200 61	225 68.6	250 76.2	275 83.8	300 91.4	400 121.9	500 152.4
25	7.6	37	65	92	120	147	175	202	230	257	277	278	278	279	280
50	15.2	47	75	102	130	157	185	212	240	267	271	272	273	276	277
75	22.9	57	85	112	140	167	195	222	250	265	266	267	269	272	274
100	30.5	67	95	122	150	177	205	232	257	260	262	263	265	269	271
125	38.1	77	105	132	160	187	215	242	253	255	257	259	261	266	269
150	45.7	87	114	142	169	197	224	245	248	251	254	256	257	263	266
175	53.3	97	124	152	179	207	234	241	245	248	250	252	254	260	264
200	61	107	134	162	189	217	233	237	241	244	247	249	251	258	262
225	68.6	117	144	172	199	225	230	234	238	241	244	246	248	255	260
250	76.2	127	154	182	209	222	227	231	235	238	241	243	246	253	258
275	83.8	136	164	191	213	219	224	229	232	236	238	241	243	251	256
300	91.4	146	174	201	211	217	222	226	230	233	236	239	241	249	254
350	106.7	166	192	200	207	212	217	222	226	229	232	235	237	245	250
400	121.9	181	190	197	203	209	214	218	222	225	228	231	233	241	247
450	137.2	180	188	195	201	206	211	215	218	222	225	227	230	238	244
500	152.4	179	186	193	198	203	208	212	215	219	222	224	227	235	241

No holding time less than 30 seconds

Length of lateral in feet (m), 8" (203.2 mm) diameter — Length of main line in feet (m), 10" (254 mm) diameter

Lateral ft	Lateral m	25 7.6	50 15.2	75 22.9	100 30.5	125 38.1	150 45.7	175 53.3	200 61	225 68.6	250 76.2	275 83.8	300 91.4	400 121.9	500 152.4
25	7.6	45	73	100	128	155	183	210	238	265	279	280	280	281	281
50	15.2	63	90	118	145	173	200	228	255	275	275	276	277	278	279
75	22.9	80	108	135	163	190	218	245	270	272	272	273	274	276	277
100	30.5	98	125	153	180	208	235	263	267	268	269	270	271	274	275
125	38.1	116	143	71	198	226	253	263	265	266	267	268	269	272	274
150	45.7	133	161	188	216	243	258	260	262	264	265	266	267	270	272
175	53.3	151	178	206	233	254	256	258	260	262	263	264	265	268	272
200	61	168	196	223	249	252	254	256	258	260	261	262	263	267	269
225	68.6	186	213	241	247	250	253	255	257	258	259	261	262	265	268
250	76.2	204	231	242	246	249	251	253	255	256	258	259	260	264	267
275	83.8	221	237	241	244	247	250	252	254	255	256	258	259	263	266
300	91.4	232	237	240	243	246	249	251	253	254	255	256	258	262	265
350	106.7	232	235	239	242	244	247	249	251	252	253	254	256	260	263
400	121.9	231	234	238	240	243	245	247	249	250	251	253	254	258	261
450	137.2	230	234	237	239	241	243	245	247	248	250	251	252	256	259
500	152.4	230	233	236	238	240	242	244	246	247	249	250	251	255	258

No holding time less than 30 seconds

FIGURE A.145 Low-pressure air test for building sewers *(continued)*.

Length of lateral in feet (m) 4" (101.6 mm) diameter		Length of main line in feet (m)								12" (304.8 mm) diameter					
		25	50	75	100	125	150	175	200	225	250	275	300	400	500
		7.6	15.2	22.9	30.5	38.1	45.7	53.3	61	68.6	76.2	83.8	91.4	121.9	152.4
25	7.6	44	84	123	163	202	242	282	321	332	333	334	334	336	336
50	15.2	48	88	128	167	207	246	286	323	324	326	327	328	331	333
75	22.9	53	92	132	172	211	251	290	316	317	319	321	323	327	329
100	30.5	57	97	136	176	216	255	295	308	311	313	316	317	323	326
125	38.1	62	101	141	180	220	260	297	301	304	308	310	312	319	323
150	45.7	66	106	145	185	224	264	290	295	299	302	305	308	315	319
175	53.3	70	110	150	189	229	268	283	289	293	297	300	303	311	316
200	61	75	114	154	194	233	271	277	283	288	292	296	299	308	313
225	68.6	79	119	158	198	238	265	272	278	283	288	291	295	304	310
250	76.2	84	123	163	202	242	259	267	273	278	283	287	291	301	308
275	83.8	88	128	167	207	244	254	262	269	274	279	283	287	298	305
300	91.4	92	132	172	211	239	249	257	264	270	275	279	283	295	302
350	106.7	101	141	180	218	231	241	249	256	262	268	272	276	289	297
400	121.9	110	150	189	210	223	233	242	249	255	261	266	270	283	292
450	137.2	119	158	189	204	216	227	235	243	249	255	260	264	278	288
500	152.4	128	166	184	198	210	221	229	237	243	249	254	259	273	283

No holding time less than 30 seconds

Length of lateral in feet (m) 6" (154.4 mm) diameter		Length of main line in feet (m)								6" (152.4 mm) diameter					
		25	50	75	100	125	150	175	200	225	250	275	300	400	500
		7.6	15.2	22.9	30.5	38.1	45.7	53.3	61	68.6	76.2	83.8	91.4	121.9	152.4
25	7.6	50	89	129	168	208	248	287	327	331	332	333	333	335	336
50	15.2	59	99	139	178	218	257	297	321	323	325	326	327	330	332
75	22.9	69	109	149	188	228	267	307	314	316	318	320	321	326	328
100	30.5	79	119	158	198	238	277	302	306	309	312	314	316	321	325
125	38.1	89	129	168	208	248	287	295	300	303	306	309	311	317	321
150	45.7	99	139	178	218	257	284	289	294	298	301	304	306	314	318
175	53.3	109	149	188	228	267	278	284	289	293	296	299	302	310	315
200	61	119	158	198	238	265	272	278	284	288	292	295	298	306	312
225	68.6	129	168	208	248	260	268	274	279	284	288	291	294	303	309
250	76.2	139	178	218	246	255	263	269	275	280	284	287	290	300	306
275	83.8	149	188	228	242	251	259	266	271	276	280	284	287	297	304
300	91.4	158	198	227	238	248	255	262	268	272	277	281	284	294	301
350	106.7	178	208	221	232	241	249	255	261	266	271	274	278	289	296
400	121.9	189	204	217	227	236	243	250	256	261	265	269	273	284	292
450	137.2	187	201	213	223	231	239	245	251	256	260	264	268	279	288
500	152.4	186	199	210	219	227	234	240	246	251	256	260	263	275	284

No holding time less than 30 seconds

ADOPTED: 1976
REVISED: 1982
REAFFIRMED: 1984

FIGURE A.145 Low-pressure air test for building sewers *(continued)*.

Minimum time, in seconds, for pressure to drop from 3½ (24 kPa) to 2½ (17 kPa) PSIG						
Length of line (feet)	(m)	3–4 in. 76.2–101.6 mm	6 in. 152.4 mm	8 in. 203.2 mm	10 in. 254 mm	12 in. 304.8 mm
25	7.6	30	30	30	30	30
50	15.2	30	30	35	55	79
75	22.9	30	30	53	83	119
100	30.5	30	40	70	110	158
125	38.1	30	50	88	138	198
150	45.7	30	59	106	165	238
175	53.3	31	69	123	193	277
200	61	35	79	141	220	317
225	68.6	40	89	158	248	340
250	76.2	44	99	176	275	340
275	83.8	48	109	194	283	340
300	91.4	53	119	211	283	340
325	99.1	57	129	227	283	340
350	106.7	62	139	227	283	340
375	114.3	66	148	227	283	340
400	121.9	70	158	227	283	362
450	137.2	79	170	227	283	407
500	152.4	88	170	227	314	452

FIGURE A.146 Sample air test table.

PVC-DWV TYPE I THERMAL EXPANSION TABLE
Chart Shows Length Change in Inches vs. Degrees Temperature Change
Coefficient of Linear Expansion: e = 2.9×10^{-5} in/in °F

Length (feet)	40°F	50°F	60°F	70°F	80°F	90°F	100°F
20	.278	.348	.418	.487	.557	.626	.696
40	.557	.696	.835	.974	1.114	1.235	1.392
60	.835	1.044	1.253	1.462	1.670	1.879	2.088
80	1.134	1.392	1.670	1.949	2.227	2.506	2.784
100	1.392	1.740	2.088	2.436	2.784	3.132	3.480

PVC-DWV TYPE I THERMAL EXPANSION TABLE (Metric)
Chart Shows Length Change in Millimeters vs. Degrees Temperature Change
Coefficient of Linear Expansion: $\frac{.2 \text{ mm}}{\text{mm °C}}$

Length (m)	4.4°C	10°C	15.6°C	21.1°C	26.7°C	32.2°C	37.8°C
6.1	7.1	8.8	10.6	12.4	14.2	15.9	17.7
12.2	14.2	17.7	21.2	24.7	28.3	31.4	35.4
18.3	21.2	26.5	31.8	37.1	42.4	47.7	53.0
24.4	28.8	35.4	42.4	49.5	56.6	63.7	70.7
30.5	35.4	44.2	53.0	61.9	70.7	79.6	88.4

FIGURE A.147 PVC-DWV type 1 thermal expansion table.

Pipe	Fittings		Maximum working pressure
	Schedule	Sizes	
160 psi (SDR 26) (1102.4 kPa)	40	½″ thru 8″ incl. (12.7 mm–203.2 mm)	160 psi–1102.4 kPa
	80	½″ thru 8″ incl. (12.7 mm–203.2 mm)	160 psi–1102.4 kPa
200 psi (SDR 21) (1378 kPa)	40	½″ thru 4″ incl. (12.7 mm–101.6 mm)	200 psi–1378 kPa
	80	½″ thru 8″ incl. (12.7 mm–203.2 mm)	200 psi–1378 kPa
250 psi (SDR 17) (1722.5 kPa)	40	½″ thru 3″ incl. (12.7 mm–76.2 mm)	250 psi–1722.5 kPa
	80	½″ thru 8″ incl. (12.7 mm–101.6 mm)	250 psi–1722.5 kPa
315 psi (SDR 13.5) (2170.4 kPa)	40	½″ thru 1½″ incl. (12.7 mm–38.1 mm)	315 psi–2170.4 kPa
	80	½″ thru 4″ incl. (12.7 mm–101.6 mm)	315 psi–2170.4 kPa
Schedule 40	40 80	½″ thru 1½″ incl. (12.7 mm–38.1 mm)	320 psi–2204.8 kPa
	40 80	2″ thru 4″ incl. (50.8 mm–101.6 mm)	220 psi–1515.8 kPa
	40	5″ thru 8″ incl.	160 psi–1102.4 kPa
Schedule 80	40	½″ thru 1½″ incl. (12.7 mm–38.1 mm)	320 psi–2204.8 kPa
	40	2″ thru 4″ incl. (50.8 mm–101.6 mm)	220 psi–1515.8 kPa
	40	5″ thru 8″ incl.	160 psi–1102.4 kPa
	80	½″ thru 4″ incl. (12.7 mm–101.6 mm)	320 psi–2204.8 kPa
	80	5″ thru 8″ incl. (127 mm–203.2 mm)	250 psi–1722.5 kPa

FIGURE A.148 Maximum working pressure.

Required sq. ft. of leaching area/100 gals septic tank capacity		Maximum septic tank size allowable	
	(m^2/L)		(liters)
20–25	(.005–.006)	7500	(28387.5)
40	(.010)	5000	(18925)
90	(.022)	3500	(13247.5)
120	(.030)	3000	(11355)

FIGURE A.149 Septic system requirements.

Because of the many variables encountered, it is not possible to set absolute values for waste/sewage flow rates for all situations. The designer should evaluate each situation and, if figures in this Table need modification, they should be made with the concurrence of the Administrative Authority.

Type of occupancy	Unit gallons (liters) per day
1. Airports	15 (56.8) per employee
	5 (18.9) per passenger
2. Auto washers	Check with equipment manufacturer
3. Bowling alleys (snack bar only)	75 (283.9) per lane
4. Camps:	
Campground with central	
comfort station	35 (132.5) per person
with flush toilets, no showers	25 (94.6) per person
Day camps (no meals served)	15 (56.8) per person
Summer and seasonal	50 (189.3) per person

FIGURE A.150 Estimated waste/sewage flow rates.

THRUST AT FITTINGS IN POUNDS AT 100 psi

Pipe size inches	90° bends	45° bends	22½° bends	Dead ends & tees
1½	415	225	115	295
2	645	350	180	455
2½	935	510	260	660
3	1,395	755	385	985
3½	1,780	962	495	1,260
4	2,295	1,245	635	1,620
5	3,500	1,900	975	2,490
6	4,950	2,710	1,385	3,550
8	8,300	4,500	2,290	5,860
10	12,800	6,900	3,540	9,050
12	18,100	9,800	5,000	12,800

THRUST AT FITTINGS IN PASCALS AT 689 kPa OF WATER PRESSURE

(mm)	90° bends	45° bends	22½° bends	Dead ends & tees
38.1	1846.8	1001.3	511.8	1312.8
50.8	2870.3	1557.5	801	2024.8
63.5	4160.8	2269.5	1157	3937
76.2	6207.8	3359.8	1713.3	4383.3
88.9	7921	4280.9	2202.8	5607
101.6	10212.8	5540.3	2815.8	7209
127	15575	8455	4338.8	11080.5
152.4	22027.5	12059.5	6163.3	15797.5
203.2	36935	20025	10190.5	26077
254	56960	30705	15753	40272.5
304.8	80545	43610	22250	56960

FIGURE A.151 Thrust at fittings in pounds and pascals.

Single family dwellings—number of bedrooms	Multiple dwelling units or apartments—one bedroom each	Other uses: maximum fixture units served	Minimum septic tank capacity in gallons (liters)
1 or 2		15	750 (2838)
3		20	1000 (3785)
4	2 units	25	1200 (4542)
5 or 6	3	33	1500 (5677.5)
	4	45	2000 (7570)
	5	55	2250 (8516.3)
	6	60	2500 (9462.5)
	7	70	2750 (10408.8)
	8	80	3000 (11355)
	9	90	3250 (12301.3)
	10	100	3500 (13247.5)

Extra bedroom, 150 gallons (567.8 liters) each.
Extra dwelling units over 10, 250 gallons (946.3 liters) each.
Extra fixture units over 100, 25 gallons (94.6 liters) per fixture units.

*Note: Septic tank sizes in this table include sludge storage capacity and the connection of domestic food waste disposal units without further volume increase.

FIGURE A.152 Capacity of septic tanks.

METRIC SYSTEM
(INTERNATIONAL SYSTEM OF UNITS – SI)
(Continued)

TO CONVERT	INTO	MULTIPLY BY
Liters	Gallons (U.S. liquid)	0.2642
Meters	Feet	3.281
Meters	Inches	39.37
Meters	Yards	1.094
Meters/second	Feet/second	3.281
Meters/second	Miles/hr	2.237
Miles (statute)	Kilometers	1.609
Miles/hour	Meters/minute	26.82
Millimeters	Inches	0.03937
Ounces (fluid)	Liters	0.02957
Pints (liquid)	Cubic centimeters	473.2
Pounds	Kilograms	0.4536
PSI	Pascals	6,895
Quarts (liquid)	Liters	0.9463
Radians	Degrees	57.30
Square inches	Square millimeters	645.2
Square meters	Square inches	1,550
Square millimeters	Square inches	1.550×10^{-3}
Watts	Btu/hour	3.4129
Watts	Btu/minute	0.05688
Watts	Foot-pounds/second	0.7378
Watts	Horsepower	1.341×10^{-3}

FIGURE A.153 Metric conversions. *(Reprinted from the 2000 Uniform Plumbing Code (UPC) with the permission of the International Association of Plumbing and Mechanical Officials (IAPMO).*

METRIC SYSTEM
(INTERNATIONAL SYSTEM OF UNITS – SI)

TO CONVERT	INTO	MULTIPLY BY
Atmospheres	Cms of mercury	76.0
Btu	Joules	1,054.8
Btu/hour	Watts	0.2931
Btu/minute	Kilowatts	0.01757
Btu/minute	Watts	17.57
Centigrade	Fahrenheit	(°C x 9/5) + 32°
Circumference	Radians	6.283
Cubic centimeters	Cubic inches	0.06102
Cubic feet	Cubic meters	0.02832
Cubic feet	Liters	28.32
Cubic feet/minute	Cubic cms/second	472.0
Cubic inches	Cubic cms	16.39
Cubic inches	Liters	0.01639
Cubic meters	Gallons (U.S. liquid)	264.2
Feet	Centimeters	30.48
Feet	Meters	0.3048
Feet	Millimeters	304.8
Feet of water	Kgs/square cm	0.03048
Foot-pounds	Joules	1.356
Foot-pounds/minute	Kilowatts	2.260×10^{-5}
Foot-pounds/second	Kilowatts	1.356×10^{-3}
Gallons	Liters	3.785
Horsepower	Watts	745.7
Horsepower-hours	Joules	2.684×10^{6}
Horsepower-hours	Kilowatt-hours	0.7457
Joules	Btu	9.480×10^{-4}
Joules	Foot-pounds	0.7376
Joules	Watt-hours	2.778×10^{-4}
Kilograms	Pounds	2.205
Kilograms	Tons (short)	1.102×10^{-3}
Kilometers	Miles	0.6214
Kilometers/hour	Miles/hour	0.6214
Kilowatts	Horsepower	1.341
Kilowatt-hours	Btu	3,413
Kilowatt-hours	Foot-pounds	2.655×10^{6}
Kilowatt-hours	Joules	3.6×10^{6}
Liters	Cubic feet	0.03531

FIGURE A.154 Metric conversions. *(Reprinted from the* 2000 Uniform Plumbing Code (UPC) *with the permission of the International Association of Plumbing and Mechanical Officials (IAPMO).*

Equivalent Length of Pipe for Various Fittings

Diameter of fitting	90° Standard Elbow	45° Standard Elbow	Standard Tee 90°	Coupling or Straight Run of Tee	Gate Valve	Globe Valve	Angle Valve
mm	mm	mm	mm	mm	mm	mm	mm
10	305	183	457	91	61	2438	1219
15	610	366	914	183	122	4572	2438
20	762	457	1219	244	152	6096	3658
25	914	549	1524	274	183	7620	4572
32	1219	732	1829	366	244	10668	5486
40	1524	914	2134	457	305	13716	6706
50	2134	1219	3048	610	396	16764	8534
65	2438	1524	3658	762	488	19812	10363
80	3048	1829	4572	914	610	24384	12192
100	4267	2438	6401	1219	823	38100	16764
125	5182	3048	7620	1524	1006	42672	21336
150	6096	3658	9144	1829	1219	50292	24384

*Allowances are based on non-recessed threaded fittings. Use one-half (1/2) the allowances for recessed threaded fittings or streamline solder fittings.

FIGURE A.155 Metric equivalent pipe lengths. *(Reprinted from the* 2000 Uniform Plumbing Code (UPC) *with the permission of the International Association of Plumbing and Mechanical Officials (IAPMO).*

Example

Fixture Units and Estimated Demands

	Building Supply Demand				Branch to Hot Water System		
Kind of Fixtures	No. of Fixtures	Fixture Unit Demand	Total Units	Building Supply Demand in gpm (L per sec)	No. of Fixtures	Fixture Unit Demand Calculation	Demand in gallons per minute (L per sec)
Water Closets	130	8.0	1040	–	–	–	–
Urinals	30	4.0	120	–	–	–	–
Shower Heads	12	2.0	24	–	12	12 x 2 x 3/4 = 18	–
Lavatories	100	1.0	100	–	100	100 x 1 x 3/4 = 75	–
Service Sinks	27	3.0	81	–	27	27 x 3 x 3/4 = 61	–
Total			1365	252 gpm (15.8 L/s)		154	55 gpm (3.4 L/s)

Allowing for 15 psi (103.4 kPa) at the highest fixture under the maximum demand of 252 gallons per minute (15.8 L/sec.), the pressure available for friction loss is found by the following:

$$55 - [15 + (45 \times 0.43)] = 20.65 \text{ psi}$$

$$\text{Metric: } 379 - [103.4 + (13.7 \times 9.8)] = 142.3 \text{ kPa}$$

The allowable friction loss per 100 feet (30.4 m) of pipe is therefore:

$$100 \times 20.65 \div 200 = 10.32 \text{ psi}$$

$$\text{Metric: } 30.4 \times 142.3 \div 60.8 = 71.1 \text{ kPa}$$

FIGURE A.156 Estimated demand can be figured with this information. *(Reprinted from the* 2000 Uniform Plumbing Code (UPC) *with the permission of the International Association of Plumbing and Mechanical Officials (IAPMO).*

°C	Base temperature	°F
	−100°–30°	
−73	−100	−148
−68	−90	−130
−62	−80	−112
−57	−70	−94
−51	−60	−76
−46	−50	−58
−40	−40	−40
−34.4	−30	−22
−28.9	−20	−4
−23.3	−10	14
−17.8	0	32
−17.2	1	33.8
−16.7	2	35.6
−16.1	3	37.4
−15.6	4	39.2
−15.0	5	41.0
−14.4	6	42.8
−13.9	7	44.6
−13.3	8	46.4
−12.8	9	48.2
−12.2	10	50.0
−11.7	11	51.8
−11.1	12	53.6
−10.6	13	55.4
−10.0	14	57.2
	31°–71°	
°C	*Base temperature*	*°F*
−0.6	31	87.8
0	32	89.6
0.6	33	91.4
1.1	34	93.2
1.7	35	95.0
2.2	36	96.8
2.8	37	98.6
3.3	38	100.4
3.9	39	102.2
4.4	40	104.0
5.0	41	105.8
5.6	42	107.6

FIGURE A.157 Temperature conversion.

Vacuum in inches of mercury	Boiling point
29	76.62
28	99.93
27	114.22
26	124.77
25	133.22
24	140.31
23	146.45
22	151.87
21	156.75
20	161.19
19	165.24
18	169.00
17	172.51
16	175.80
15	178.91
14	181.82
13	184.61
12	187.21
11	189.75
10	192.19
9	194.50
8	196.73
7	198.87
6	200.96
5	202.25
4	204.85
3	206.70
2	208.50
1	210.25

FIGURE A.158 Boiling points of water based on pressure.

LOAD VALUES ASSIGNED TO FIXTURES[a]

FIXTURE	OCCUPANCY	TYPE OF SUPPLY CONTROL	LOAD VALUES, IN WATER SUPPLY FIXTURE UNITS (wsfu)		
			Cold	Hot	Total
Bathroom group	Private	Flush tank	2.7	1.5	3.6
Bathroom group	Private	Flush valve	6.0	3.0	8.0
Bathtub	Private	Faucet	1.0	1.0	1.4
Bathtub	Public	Faucet	3.0	3.0	4.0
Bidet	Private	Faucet	1.5	1.5	2.0
Combination fixture	Private	Faucet	2.25	2.25	3.0
Dishwashing machine	Private	Automatic		1.4	1.4
Drinking fountain	Offices, etc.	$^3/_8$" valve	0.25		0.25
Kitchen sink	Private	Faucet	1.0	1.0	1.4
Kitchen sink	Hotel, restaurant	Faucet	3.0	3.0	4.0
Laundry trays (1 to 3)	Private	Faucet	1.0	1.0	1.4
Lavatory	Private	Faucet	0.5	0.5	0.7
Lavatory	Public	Faucet	1.5	1.5	2.0
Service sink	Offices, etc.	Faucet	2.25	2.25	3.0
Shower head	Public	Mixing valve	3.0	3.0	4.0
Shower head	Private	Mixing valve	1.0	1.0	1.4
Urinal	Public	1" flush valve	10.0		10.0
Urinal	Public	$^3/_4$" flush valve	5.0		5.0
Urinal	Public	Flush tank	3.0		3.0
Washing machine (8 lbs.)	Private	Automatic	1.0	1.0	1.4
Washing machine (8 lbs.)	Public	Automatic	2.25	2.25	3.0
Washing machine (15 lbs.)	Public	Automatic	3.0	3.0	4.0
Water closet	Private	Flush valve	6.0		6.0
Water closet	Private	Flush tank	2.2		2.2
Water closet	Public	Flush valve	10.0		10.0
Water closet	Public	Flush tank	5.0		5.0
Water closet	Public or private	Flushometer tank	2.0		2.0

For SI: 1 inch = 25.4 mm, 1 pound = 0.454 kg.

a. For fixtures not listed, loads should be assumed by comparing the fixture to one listed using water in similar quantities and at similar rates. The assigned loads for fixtures with both hot and cold water supplies are given for separate hot and cold water loads and for total load, the separate hot and cold water loads being three-fourths of the total load for the fixture in each case.

FIGURE A.159 Every fixture involved in plumbing has a load value. They are determined here. *(Courtesy of International Code Council, Inc. and* International Plumbing Code 2000*)*

Table for Estimating Demand

SUPPLY SYSTEMS PREDOMINANTLY FOR FLUSH TANKS			SUPPLY SYSTEMS PREDOMINANTLY FOR FLUSH VALVES		
Load	Demand		Load	Demand	
(Water supply fixture units)	(Gallons per minute)	(Cubic feet per minute)	(Water supply fixture units)	(Gallons per minute)	(Cubic feet per minute)
1	3.0	0.04104			
2	5.0	0.0684			
3	6.5	0.86892			
4	8.0	1.06944			
5	9.4	1.256592	5	15.0	2.0052
6	10.7	1.430376	6	17.4	2.326032
7	11.8	1.577424	7	19.8	2.646364
8	12.8	1.711104	8	22.2	2.967696
9	13.7	1.831416	9	24.6	3.288528
10	14.6	1.951728	10	27.0	3.60936
11	15.4	2.058672	11	27.8	3.716304
12	16.0	2.13888	12	28.6	3.823248
13	16.5	2.20572	13	29.4	3.930192
14	17.0	2.27256	14	30.2	4.037136
15	17.5	2.3394	15	31.0	4.14408
16	18.0	2.90624	16	31.8	4.241024
17	18.4	2.459712	17	32.6	4.357968
18	18.8	2.513184	18	33.4	4.464912
19	19.2	2.566656	19	34.2	4.571856
20	19.6	2.620128	20	35.0	4.6788
25	21.5	2.87412	25	38.0	5.07984
30	23.3	3.114744	30	42.0	5.61356
35	24.9	3.328632	35	44.0	5.88192
40	26.3	3.515784	40	46.0	6.14928
45	27.7	3.702936	45	48.0	6.41664
50	29.1	3.890088	50	50.0	6.684

FIGURE A.160 This table will let a user estimate demand. *(Courtesy of International Code Council, Inc. and* International Plumbing Code 2000*)*

Table for Estimating Demand—cont'd

SUPPLY SYSTEMS PREDOMINANTLY FOR FLUSH TANKS			SUPPLY SYSTEMS PREDOMINANTLY FOR FLUSH VALVES		
Load	Demand		Load	Demand	
(Water supply fixture units)	(Gallons per minute)	(Cubic feet per minute)	(Water supply fixture units)	(Gallons per minute)	(Cubic feet per minute)
60	32.0	4.27776	60	54.0	7.21872
70	35.0	4.6788	70	58.0	7.75344
80	38.0	5.07984	80	61.2	8.181216
90	41.0	5.48088	90	64.3	8.595624
100	43.5	5.81508	100	67.5	9.0234
120	48.0	6.41664	120	73.0	9.75864
140	52.5	7.0182	140	77.0	10.29336
160	57.0	7.61976	160	81.0	10.82808
180	61.0	8.15448	180	85.5	11.42964
200	65.0	8.6892	200	90.0	12.0312
225	70.0	9.3576	225	95.5	12.76644
250	75.0	10.0260	250	101.0	13.50168
275	80.0	10.6944	275	104.5	13.96956
300	85.0	11.3628	300	108.0	14.43744
400	105.0	14.0364	400	127.0	16.97736
500	124.0	16.57632	500	143.0	19.11624
750	170.0	22.7256	750	177.0	23.66136
1,000	208.0	27.80544	1,000	208.0	27.80544
1,250	239.0	31.94952	1,250	239.0	31.94952
1,500	269.0	35.95992	1,500	269.0	35.95992
1,750	297.0	39.70296	1,750	297.0	39.70296
2,000	325.0	43.446	2,000	325.0	43.446
2,500	380.0	50.7984	2,500	380.0	50.7984
3,000	433.0	57.88344	3,000	433.0	57.88344
4,000	535.0	70.182	4,000	525.0	70.182
5,000	593.0	79.27224	5,000	593.0	79.27224

For SI: 1 gpm = 3.785 L/m, 1 cfm = 0.4719 L/s.

FIGURE A.161 The table for estimating demand for flush tanks and valves. *(Courtesy of International Code Council, Inc. and* International Plumbing Code 2000*)*

LOSS OF PRESSURE THROUGH TAPS AND TEES IN POUNDS PER SQUARE INCH (psi)

GALLONS PER MINUTE	SIZE OF TAP OR TEE (inches)						
	5/8	3/4	1	1 1/4	1 1/2	2	3
10	1.35	0.64	0.18	0.08			
20	5.38	2.54	0.77	0.31	0.14		
30	12.1	5.72	1.62	0.69	0.33	0.10	
40		10.2	3.07	1.23	0.58	0.18	
50		15.9	4.49	1.92	0.91	0.28	
60			6.46	2.76	1.31	0.40	
70			8.79	3.76	1.78	0.55	0.10
80			11.5	4.90	2.32	0.72	0.13
90			14.5	6.21	2.94	0.91	0.16
100			17.94	7.67	3.63	1.12	0.21
120			25.8	11.0	5.23	1.61	0.30
140			35.2	15.0	7.12	2.20	0.41
150				17.2	8.16	2.52	0.47
160				19.6	9.30	2.92	0.54
180				24.8	11.8	3.62	0.68
200				30.7	14.5	4.48	0.84
225				38.8	18.4	5.6	1.06
250				47.9	22.7	7.00	1.31
275					27.4	7.70	1.59
300					32.6	10.1	1.88

For SI: 1 inch = 25.4 mm, 1 psi = 6.895 kPa, 1 gpm = 3.785 L/m.

FIGURE A.162 Pressure can be lost in taps and tees. This examines the numbers. *(Courtesy of International Code Council, Inc. and* International Plumbing Code 2000*)*

ALLOWANCE IN EQUIVALENT LENGTH OF PIPE FOR FRICTION LOSS IN VALVES AND THREADED FITTINGS (feet)

FITTING OR VALVE	PIPE SIZES (inches)							
	1/2	3/4	1	1 1/4	1 1/2	2	2 1/2	3
45-degree elbow	1.2	1.5	1.8	2.4	3.0	4.0	5.0	6.0
90-degree elbow	2.0	2.5	3.0	4.0	5.0	7.0	8.0	10.0
Tee, run	0.6	0.8	0.9	1.2	1.5	2.0	2.5	3.0
Tee, branch	3.0	4.0	5.0	6.0	7.0	10.0	12.0	15.0
Gate valve	0.4	0.5	0.6	0.8	1.0	1.3	1.6	2.0
Balancing valve	0.8	1.1	1.5	1.9	2.2	3.0	3.7	4.5
Plug-type cock	0.8	1.1	1.5	1.9	2.2	3.0	3.7	4.5
Check valve, swing	5.6	8.4	11.2	14.0	16.8	22.4	28.0	33.6
Globe valve	15.0	20.0	25.0	35.0	45.0	55.0	65.0	80.0
Angle valve	8.0	12.0	15.0	18.0	22.0	28.0	34.0	40.0

For SI: 1 inch = 25.4 mm, 1 foot = 304.8 mm, 1 degree = 0.0175 rad.

FIGURE A.163 This chart examines the allowances involved in friction loss in valves and threaded fittings. *(Courtesy of International Code Council, Inc. and* International Plumbing Code 2000*)*

PRESSURE LOSS IN FITTINGS AND VALVES EXPRESSED AS EQUIVALENT LENGTH OF TUBE[a] (feet)

NOMINAL OR STANDARD SIZE (inches)	FITTINGS					VALVES			
	Standard Ell		90-Degree Tee						
	90 Degree	45 Degree	Side Branch	Straight Run	Coupling	Ball	Gate	Butterfly	Check
3/8	0.5	—	1.5	—	—	—	—	—	1.5
1/2	1	0.5	2	—	—	—	—	—	2
5/8	1.5	0.5	2	—	—	—	—	—	2.5
3/4	2	0.5	3	—	—	—	—	—	3
1	2.5	1	4.5	—	—	0.5	—	—	4.5
1 1/4	3	1	5.5	0.5	0.5	0.5	—	—	5.5
1 1/2	4	1.5	7	0.5	0.5	0.5	—	—	6.5
2	5.5	2	9	0.5	0.5	0.5	0.5	7.5	9
2 1/2	7	2.5	12	0.5	0.5	—	1	10	11.5
3	9	3.5	15	1	1	—	1.5	15.5	14.5
3 1/2	9	3.5	14	1	1	—	2	—	12.5
4	12.5	5	21	1	1	—	2	16	18.5
5	16	6	27	1.5	1.5	—	3	11.5	23.5
6	19	7	34	2	2	—	3.5	13.5	26.5
8	29	11	50	3	3	—	5	12.5	39

For SI: 1 inch = 25.4 mm, 1 foot = 304.8 mm, 1 degree = 0.0175 rad.

a. Allowances are for streamlined soldered fittings and recessed threaded fittings. For threaded fittings, double the allowances shown in the table. The equivalent lengths presented above are based on a C factor of 150 in the Hazen-Williams friction loss formula. The lengths shown are rounded to the nearest half-foot.

FIGURE A.164 You can determine pressure losses as equivalent lengths from this table. *(Courtesy of International Code Council, Inc. and* International Plumbing Code 2000*)*

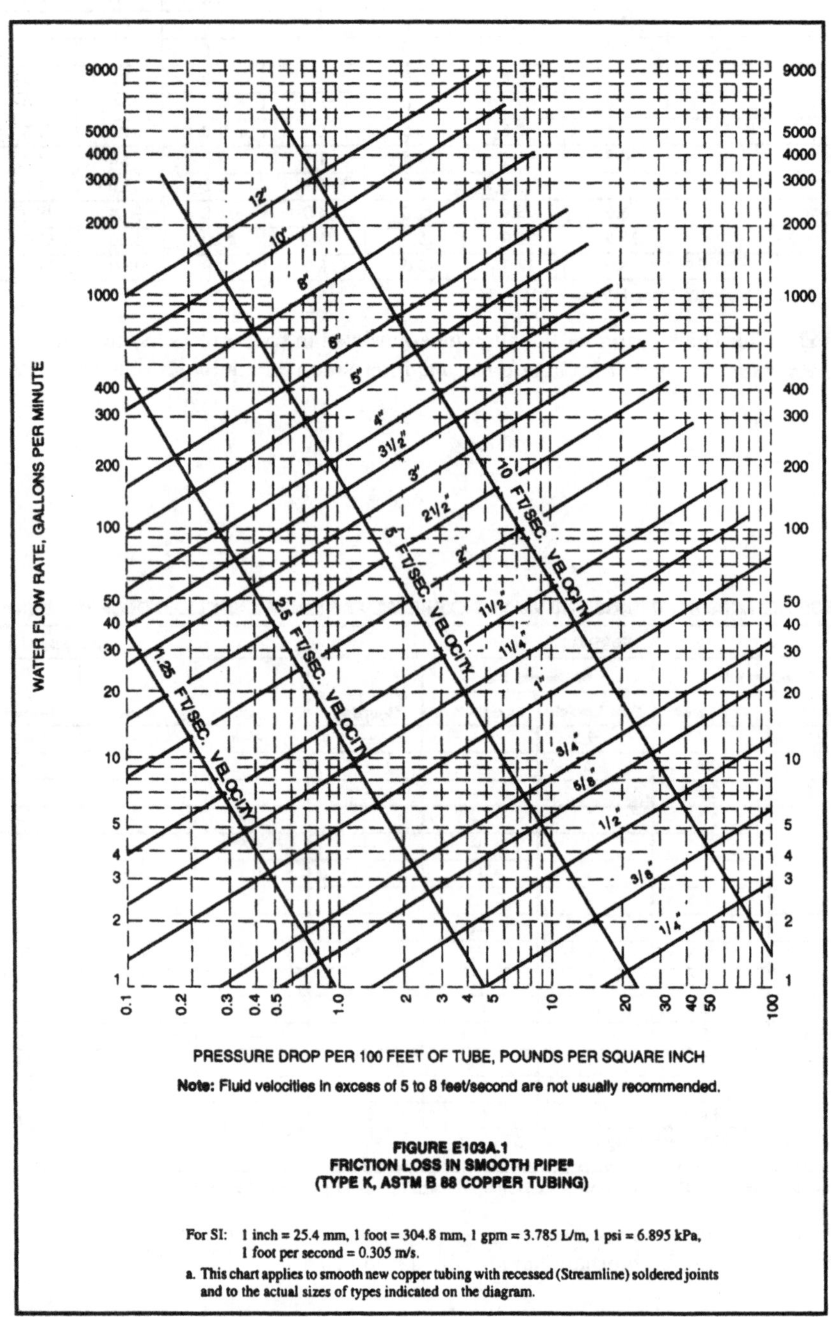

FIGURE A.165 This is one of several tables that determines friction loss. *(Courtesy of International Code Council, Inc. and* International Plumbing Code 2000*)*

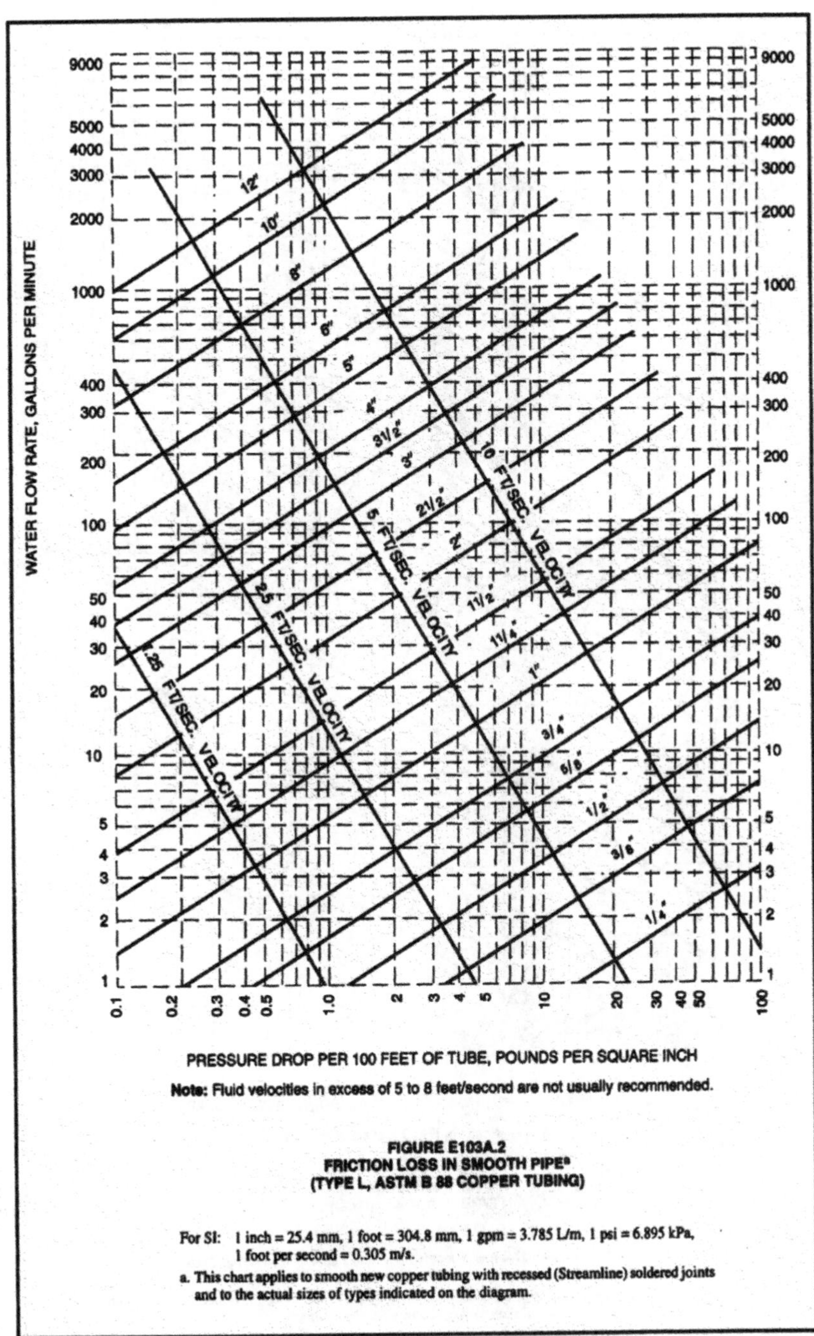

FIGURE E103A.2
FRICTION LOSS IN SMOOTH PIPE[a]
(TYPE L, ASTM B 88 COPPER TUBING)

For SI: 1 inch = 25.4 mm, 1 foot = 304.8 mm, 1 gpm = 3.785 L/m, 1 psi = 6.895 kPa, 1 foot per second = 0.305 m/s.

a. This chart applies to smooth new copper tubing with recessed (Streamline) soldered joints and to the actual sizes of types indicated on the diagram.

FIGURE A.166 This is one of several tables that determines friction loss. *(Courtesy of International Code Council, Inc. and* International Plumbing Code 2000*)*

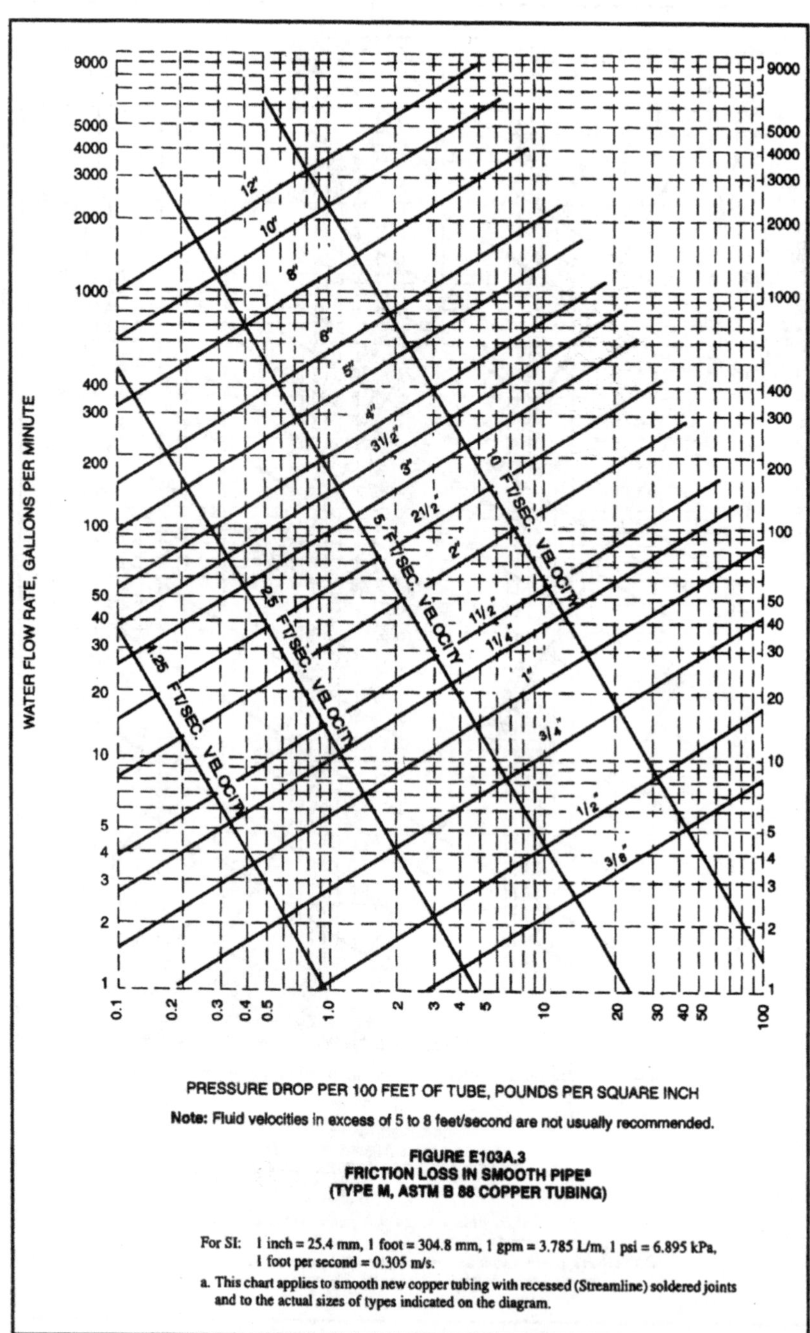

FIGURE A.167 This is one of several tables that determines friction loss. *(Courtesy of International Code Council, Inc. and* International Plumbing Code 2000*)*

Allowance in equivalent length of pipe for friction loss in valves and threaded fittings.*

Equivalent Length of Pipe for Various Fittings

Diameter of fitting **Inches**	90° Standard Elbow **Feet**	45° Standard Elbow **Feet**	Standard Tee 90° **Feet**	Coupling or Straight Run of Tee **Feet**	Gate Valve **Feet**	Globe Valve **Feet**	Angle Valve **Feet**
3/8	1.0	0.6	1.5	0.3	0.2	8	4
1/2	2.0	1.2	3.0	0.6	0.4	15	8
3/4	2.5	1.5	4.0	0.8	0.5	20	12
1	3.0	1.8	5.0	0.9	0.6	25	15
1-1/4	4.0	2.4	6.0	1.2	0.8	35	18
1-1/2	5.0	3.0	7.0	1.5	1.0	45	22
2	7.0	4.0	10.0	2.0	1.3	55	28
2-1/2	8.0	5.0	12.0	2.5	1.6	65	34
3	10.0	6.0	15.0	3.0	2.0	80	40
4	14.0	8.0	21.0	4.0	2.7	125	55
5	17.0	10.0	25.0	5.0	3.3	140	70
6	20.0	12.0	30.0	6.0	4.0	165	80

FIGURE A.168 Equivalent pipe length of various fittings. *(Reprinted from the* 2000 Uniform Plumbing Code (UPC) *with the permission of the International Association of Plumbing and Mechanical Officials (IAPMO))*

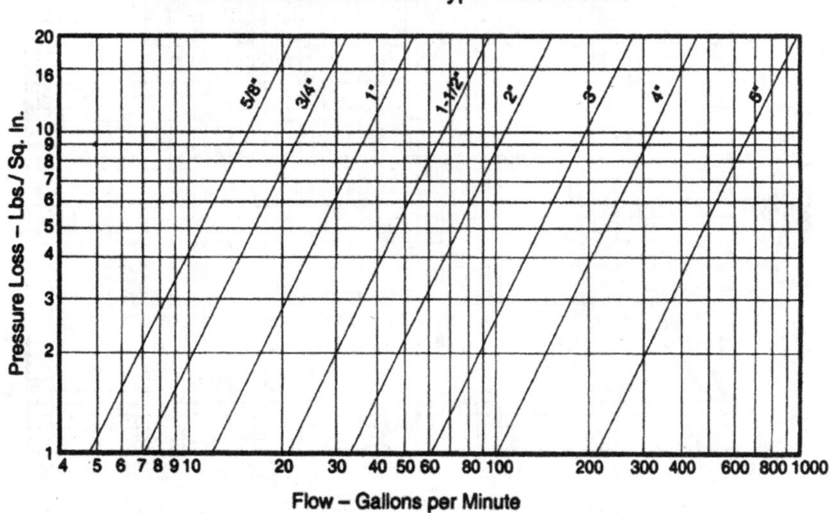

FIGURE A.169 English and metric unit information about friction loss. *(Reprinted from the* 2000 Uniform Plumbing Code (UPC) *with the permission of the International Association of Plumbing and Mechanical Officials (IAPMO))*

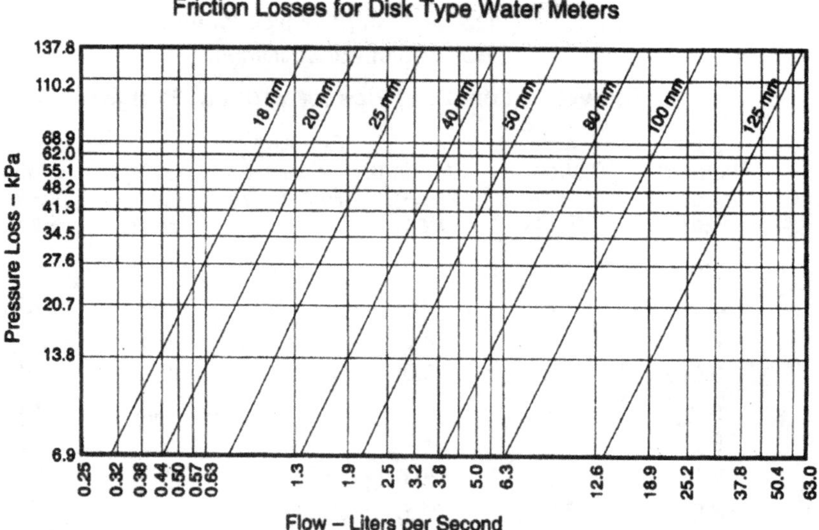

FIGURE A.170 English and metric unit information about friction loss. *(Reprinted from the* 2000 Uniform Plumbing Code (UPC) *with the permission of the International Association of Plumbing and Mechanical Officials (IAPMO))*

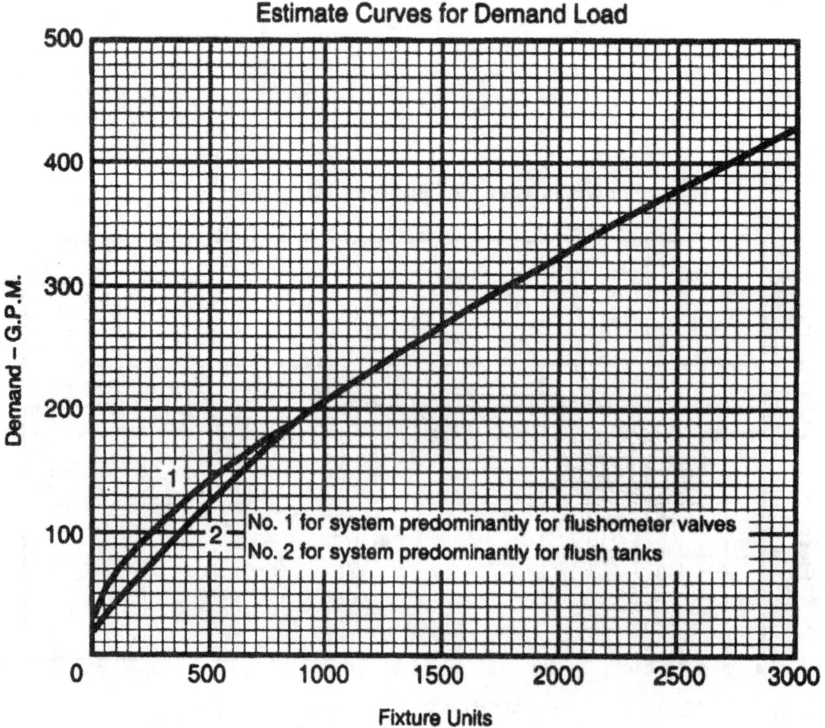

FIGURE A.171 English and metric lengths. *(Reprinted from the* 2000 Uniform Plumbing Code (UPC) *with the permission of the International Association of Plumbing and Mechanical Officials (IAPMO))*

ANSWERS AND REFERENCES

Code references to the *IPC* are to the *International Plumbing Code.*

Code references to the *UPC* are to the *Uniform Plumbing Code.*

	Code Reference	Correct Answer
CHAPTER 1		
1.	*IPC* 102.8, *UPC* 101.4	d
2.	*UPC* 102.9	c
3.	*IPC* 106.2	a
4.	*IPC* 105.2	d
5.	*IPC* 107.1	c
6.	*IPC* 106.5	c
7.	*IPC* 106.1	a
8.	*IPC* 106.5.8	c
9.	*IPC* 108.5	d
10.	*IPC* 107.5	b
11.	*IPC* 106.5.4, *UPC* 103.3.3	d
12.	*IPC* 106.5.3, *UPC* 103.3.3	b
13.	*IPC* 106.5.6, *UPC* 103.3.1	b
14.	*IPC* 109.1, *UPC* 103	d
15.	*IPC* 110, *UPC* 103.6.2	a
16.	*IPC* 106.3, *UPC* 103.2	b
17.	*IPC* 107.2, *UPC* 103.5	c
18.	*IPC* 103.2, *UPC* 102.1	d
19.	No code reference	a
20.	*IPC* 108.2	b
21.	*IPC* 106.2	d
22.	*IPC* 102.8	b
23.	*IPC* 107.6	d
24.	*IPC* 101.2	a
25.	*IPC* 101.2	a
26.	*IPC* 106.1	c
27.	No code reference	c
28.	*IPC* 103	c
29.	*IPC* 101.2	c
30.	*IPC* 103	b
31.	*IPC* 301.7	d
32.	*IPC* 311	b
33.	*IPC* 307.1	a
34.	*IPC* 104.4	a
CHAPTER 2		
1.	*IPC* 103.4	d
2.	*IPC* 314.2	b
3.	*IPC* 305	d
4.	*IPC* 305.3	a
5.	*IPC* 307.6	b
6.	*IPC* 305.6	d
7.	*IPC* 105.6	b
8.	*IPC* 306.2	d
9.	*IPC* 306.3	c
10.	*IPC* 307.2	d
11.	*IPC* 315	b
12.	*IPC* 403.2	c
13.	*IPC* 301.3	d
14.	*IPC* 307.5	b
15.	*IPC* 1003.4	d
16.	*IPC* 1003.4	a
17.	*IPC* 305	d
18.	*IPC* 306.2.3	d

	Code Reference	Correct Answer
19.	*IPC* 306. 3	b
20.	*IPC* 305.5	d
21.	*IPC* 107.1	a
22.	*IPC* 307.1	b
23.	*IPC* 307.3.9	c
24.	*IPC* 312.1.1	d
25.	*IPC* 301.1	a
26.	*IPC* 305.4	c
27.	*IPC* 316	d
28.	*IPC* 312.2	b
29.	*IPC* 312.1	b
30.	*IPC* 312.1	a
31.	*IPC* 312.4	c
32.	*IPC* 312.2	b
33.	*IPC* 312.5	c
34.	*IPC* 312.3	a

CHAPTER 3

	Code Reference	Correct Answer
1.	*IPC* 602.3.1	a
2.	*IPC* 602.3.1	d
3.	*IPC* definitions	c
4.	*IPC* 606.1	d
5.	*IPC* 606.1	b
6.	*IPC* 604.8	a
7.	*IPC* 604.5	c
8.	*IPC* 607.1	d
9.	*IPC* 607.1	d
10.	*IPC* 607.1	a
11.	*IPC* 602.2	c
12.	*IPC* 602.2	d
13.	*IPC* 602.2	b
14.	*IPC* 603.1	d
15.	*IPC* 603.2	b
16.	*IPC* 603.2.1	a
17.	*IPC* 603.2.1	a
18.	*IPC* 604.9	d
19.	*IPC* 604.6	a
20.	*IPC* 604.10.3	b
21.	*IPC* 608.17.5	c
22.	*IPC* 605.9	c
23.	*IPC* per Table 3.1	a
24.	*IPC* per Table 3.1	b
25.	*IPC* per Table 3.1	c
26.	*IPC* per Table 3.1	b
27.	*IPC* per Table 3.1	a
28.	*IPC* per Table 3.1	b
29.	*IPC* 608.8.2	b
30.	*IPC* 608.13.5	d
31.	*IPC* 608.16.5	a
32.	*IPC* 609.4	c
33.	*IPC* 609.13.1	c
34.	*IPC* 609.14.4	d
35.	*IPC* 606.5.4	a
36.	*IPC* 607.4	b
37.	*IPC* 608.16.1	c
38.	*IPC* 610.1.2	c
39.	*IPC* 610.1.2, *UPC* 609	d
40.	*IPC* 608.13.4	b
41.	*IPC* Table 308.5	b
42.	*IPC* Table 308.5	c
43.	*IPC* Table 602.3.5.1	a
44.	*IPC* Table 606.4	c

	Code Reference	Correct Answer
45.	*IPC* 605.22.3, *UPC* 605.4.3	b
46.	*IPC* Table 608.3.1	d
47.	*IPC* 603.3.1, *UPC* 605.3.3	c
48.	*IPC* 605.5.1, *UPC* 605.3.3.1	a
49.	*UPC* 608.3, *IPC* 604.8	d
50.	*UPC* 610, 12	c

CHAPTER 4

	Code Reference	Correct Answer
1.	*IPC* 501.6	d
2.	*IPC* 501.2	d
3.	*IPC* 502.3	b
4.	*IPC* 502.3	a
5.	*IPC* 502.5	c
6.	*IPC* 504.7	b
7.	*IPC* 504.6	c
8.	*IPC* 504.6	c
9.	*IPC* 501.7	d
10.	*IPC* 503.1	a
11.	*IPC* 502.1	d
12.	*IPC* 502.2	a
13.	*IPC* 502.4	b
14.	*IPC* 504.1	c
15.	*IPC* 504.4	a
16.	*IPC* 504.4	b
17.	*IPC* 504.6	d
18.	*IPC* 501.5	d
19.	*IPC* 503.1	c
20.	*IPC* 608.16.3, *UPC* 603.4.10	a
21.	*IPC* 608.16.2	c
22.	*UPC* 608.16.2	a
23.	*IPC* 608.16.3, *UPC* 603.4.4.1	a
24.	*UPC* 505.1	b
25.	*UPC*.801.6, *IPC* 802.1.3	d
26.	*UPC* 504.7.2	a
27.	*IPC* 504.5	d
28.	*IPC* 504.6	b
29.	*IPC* 504.7	a
30.	*IPC* 504.4	a

CHAPTER 5

	Code Reference	Correct Answer
1.	*IPC* 708.3.1	d
2.	*IPC* 708.3.2	a
3.	*IPC* 708.3.3	c
4.	*IPC* 702.6	c
5.	*IPC* 708.3.3	c
6.	*IPC* 708.7	c
7.	*IPC* 708.7	a
8.	*IPC* 708.8	c
9.	*IPC* 709.2	a
10.	*IPC* 709.3, *UPC* 702.3	a
11.	*IPC* 711.2	a
12.	*IPC* 712.3.2	b
13.	*IPC* 712.3.2	c
14.	*IPC* 712.3.4	b
15.	*IPC* 712.3.5	d
16.	*IPC* 712.4.2	b
17.	*IPC* 712.4.2	c
18.	*IPC* 712.4.2	c
19.	*IPC* 713.3	d
20.	*IPC* 713.7.2	c
21.	*IPC* 713.9	b
22.	*IPC* 713.9.1	d

	Code Reference	Correct Answer
23.	*IPC* 713.9.3	c
24.	*IPC* 713.11.1	b
25.	*IPC* 713.11.2	c
26.	*IPC* 713.11.3	d
27.	*IPC* 704.1	a
28.	*IPC* Table 701.7	d
29.	*IPC* Table 706.3	c
30.	*IPC* Table 5.1	b
31.	*IPC* per Table 5.1	b
32.	*IPC* per Table 5.1	c
33.	*IPC* per Table 5.4	c
34.	*IPC* per Table 5.3	a
35.	*IPC* 715	b
36.	*IPC* per Table 5.3	a
37.	*IPC* per Table 5.2	c
38.	*UPC* definitions	b
39.	*IPC* Table 709.1	c
40.	*IPC* 412.4	c
41.	*IPC* Table 710.1(1)	b
42.	*IPC* 312.7	b
43.	*UPC* 710.3	a
44.	*UPC* 711.1	d
45.	*UPC* 713.4	c

CHAPTER 6

	Code Reference	Correct Answer
1.	*IPC* Chapter 2	b
2.	*IPC* Chapter 2	c
3.	*IPC* 405.3.1	d
4.	*IPC* 421	d
5.	*IPC* 403.3	c
6.	*IPC* 403.3	b
7.	*IPC* 403.2	c
8.	*IPC* 403.1.1	a
9.	*IPC* 403.3	a
10.	*IPC* 403.3	d
11.	*IPC* 403.3.3	d
12.	*IPC* 405.3.1	a
13.	*IPC* 405.3.1	c
14.	*IPC* 405.3.1	b
15.	*IPC* 424.3	c
16.	*IPC* 424.3	b
17.	*IPC* 405.3.1	d
18.	*IPC* 407.2	c
19.	*IPC* 405.1	a
20.	*IPC* 411.2, *UPC* 416.5	d
21.	*IPC* 410.4	c
22.	*IPC* 413.3	a
23.	*IPC* 409.2	b
24.	*IPC* 409.2, 114.9	b
25.	*IPC* 409.2, 114.9	d
26.	*IPC* 419.2	c
27.	*IPC* 401.3	d
28.	*IPC* 713.11.2	b
29.	*IPC* 417.3	c
30.	*IPC* definitions	b
31.	*UPC* 403.2, *IPC* 604.4	a
32.	*UPC* 402.6.3, *IPC* 420.4	d
33.	*UPC* 408.4	b
34.	*IPC* 417.3	a

	Code Reference	Correct Answer
CHAPTER 7		
1.	*UPC* 802.1.1	c
2.	*UPC* 801.1	d
3.	*UPC* .804.1	a
4.	*IPC* 802.2	d
5.	*UPC* 801.3	a
6.	*IPC* 802.1.8, *UPC* 801.4	d
7.	*IPC* 802.1.1	c
8.	*IPC* 802.2	b
9.	*UPC* 804.1	d
10.	*IPC* 802.3	a
11.	*IPC* 802.4	d
12.	*IPC* 802.1, *UPC* 801.5	c
13.	*IPC* 803.3	b
14.	*IPC* 802.1.2	d
15.	*IPC* 802.1.2	a
16.	*IPC* 802.3	d
17.	*UPC* 803	d
18.	*UPC* 807	a
19.	*IPC* 802.3	c
20.	*UPC* 804, *IPC*, 802.2	d
21.	*UPC* 802.1.4	c
22.	*UPC* 802.2.1, 802.2.2	c
23.	*UPC* 804.1, *IPC* 802.3.1	b
24.	*UPC* 801.2.1, 802.2.2	a
25.	*UPC* 801.5	d
26.	*UPC* 801.5	b
27.	*UPC* 810.1, *IPC* 803.1	c
28.	*UPC* 814.1	b
29.	*UPC* 814.3	d
30.	*UPC* 801.4	c
CHAPTER 8		
1.	*IPC* 901.3	a
2.	*IPC* 901.3	c
3.	*IPC* 904.2	d
4.	*IPC* 910.1	c
5.	*IPC* 913.3	b
6.	*IPC* 910.1	c
7.	*IPC* 901.2.3	a
8.	*IPC* 906.2	c
9.	*IPC* 906.1	b
10.	*IPC* 906.3	c
11.	*IPC* 901.3	d
12.	*IPC* 911.1	c
13.	*IPC* 906.2	a
14.	*IPC* 906.3	c
15.	*IPC* 908.1	d
16.	*IPC* 904.3	b
17.	*IPC* 902.1, *UPC* 901.2	a
18.	*IPC* 903.1	d
19.	*IPC* 903.5	b
20.	*IPC* 904.4	a
21.	*IPC* 904.4	d
22.	*IPC* 903.7	b
23.	*IPC* 903.6	c
24.	*IPC* 904.4	d
25.	*IPC* 904.4	b
26.	*IPC* 905.4	b
27.	*IPC* 905.2	a
28.	*IPC* 909.2	c

	Code Reference	Correct Answer
29.	*IPC* 909.2	c
30.	*IPC* 912.1.1	a
31.	*IPC* 907.3	d
32.	*IPC* 902.3	b
33.	*IPC* 902.2	a
34.	*IPC* 916.3	d
35.	*IPC* 916.3	b
36.	*IPC* 913.3	b
37.	*IPC* 913.3	b
38.	*IPC* 912.2.1	c
39.	*IPC* 912.2.1	c
40.	*IPC* 905.5	b
41.	*IPC* 905.6	d
42.	*IPC* 915.2.2	b
43.	*IPC* 908.1	a
44.	*IPC* 916.1	c
45.	*IPC* 915.1	d
46.	*IPC* 908.1	c
47.	*IPC* 909.1	d
48.	*IPC* 909.1	b
49.	*IPC* 917.2	a
50.	*IPC* 917.3	b
51.	*IPC* 917.4.1	c
52.	*IPC* 917.5	b
53.	*IPC* 917.4.2	a
54.	*IPC* 917.8	b
55.	*IPC* 918.3.1	d
56.	*IPC* 918.4	c
57.	*IPC* 918.4	a
58.	*IPC* 713.11.3	c
59.	*UPC* 903.2.1	d
60.	*UPC* 703.1, *IPC* 905.4	b
61.	*UPC* 903.1, *IPC* 906.1	a
62.	*IPC* 903.2	c
63.	*UPC* 906.7	d
64.	*UPC* 907.2	b

CHAPTER 9

	Code Reference	Correct Answer
1.	*IPC* 1101.2	a
2.	*IPC* 1114.1.2	d
3.	*IPC* 1101.2	b
4.	*IPC* 1113.1	d
5.	*IPC* 1102.7	d
6.	*IPC* 1104.2	a
7.	*IPC* 1102.7	c
8.	*IPC* 1104.1	d
9.	*IPC* 1101.2	c
10.	*IPC* 1101.2	d
11.	*IPC* 1106.2	b
12.	*IPC* 1113.1	b
13.	*IPC* 1114.1.2	c
14.	*IPC* 1101.2	d
15.	*IPC* 1112.1	c
16.	*IPC* 1101.2	d
17.	*IPC* 1106.1	c
18.	*IPC* 1106.5	a
19.	*IPC* 1106.3	b
20.	*IPC* 1106.4	d
21.	*IPC* 1108.1	a
22.	*IPC* 1108.2	c
23.	*IPC* 1110.1	d
24.	*IPC* 1112.1	b

	Code Reference	Correct Answer
25.	*IPC* 1111	a
26.	*UPC* 1107.1	b

CHAPTER 10

	Code Reference	Correct Answer
1.	*IPC* 708.3.2	d
2.	*IPC* 803.1	d
3.	*IPC* 803.3	a
4.	*IPC* 803.1	b
5.	*IPC* 1002.2	c
6.	*IPC* 1002.2	d
7.	*IPC* 1002.2	b
8.	*IPC* 1002.1	c
9.	*IPC* 1002.3	b
10.	*IPC* 1002.6	a
11.	*IPC* 1002.4	c
12.	*IPC* 405.8	d
13.	*IPC* 1002.1	b
14.	*IPC* 411.2	d
15.	*IPC* 411.1	a
16.	*IPC* 411.1	c
17.	*IPC* 708.3.2	c
18.	*IPC* 708.2	d
19.	*IPC* 708.8	b
20.	*IPC* 708.7	a
21.	*IPC* 1002.6	d
22.	*IPC* 1002.1	b
23.	*IPC* 1002.1	c
24.	*IPC* 708.3.2	d
25.	*IPC* 802.2	b
26.	*IPC* 1002.4	d

CHAPTER 11

	Code Reference	Correct Answer
1.	*IPC* 1003.1	b
2.	*IPC* 1003.3.4.2	d
3.	*IPC* 1003.4.2.1, *UPC* 1017.1	a
4.	*IPC* 1003.4.2.1	c
5.	*IPC* 1003.4.2.2	b
6.	*IPC* 1003.4.2.2	d
7.	*IPC* 1003.5	c
8.	*IPC* 1003.5, *UPC* 1009.3	c
9.	*IPC* 1003.6	b
10.	*IPC* 1003.6	a
11.	*IPC* 1003.7	b
12.	*IPC* 1003.8	d
13.	*IPC* 1003.8	d
14.	*IPC* 1003.10	d
15.	*IPC* 1002.1	c
16.	*IPC* 1002.1	a
17.	*IPC* 1002.1	a
18.	*IPC* 1002.1	a
19.	*IPC* 1002.3	b
20.	*IPC* 1002.1	a
21.	*IPC* 1003.3.1	d
22.	*IPC* 1003.3.1	d
23.	*IPC* 1003.3.4.2	c
24.	*IPC* 1002.9	b
25.	*IPC* 1003.4.2.1	a

CHAPTER 12

	Code Reference	Correct Answer
1.	*IPC* 1202.1	d

	Code Reference	Correct Answer
CHAPTER 13		
1.	*IPC* 917.2	a
2.	*IPC* 917.4.1	c
3.	*IPC* 917.4.1	d
4.	*IPC* 917.4.1	b
5.	*IPC* 917.4.2	c
6.	*IPC* 917.5	b
7.	*IPC* 917.5	a
8.	*IPC* 917.8	d
9.	*IPC* per Tables 13.1 and 13.2	c
10.	*IPC* per Tables 13.1 and 13.2	b
CHAPTER 14		
1.	*IPC* 1301.7	d
2.	*IPC* 1301.1	c
3.	*IPC* 1301.8, 9, 11	b
4.	*IPC* 1303.1	c
5.	*IPC* 1302.4	a
6.	*IPC* 1302.2	d
7.	*IPC* 1303.6	c
8.	*IPC* 1303.4	c
9.	*IPC* 1303.10	a
10.	*IPC* 1303.10	b
11.	*IPC* 1303.9.2	d
12.	*IPC* 1303.9.2	c
13.	*IPC* 1303.9.3	a
14.	*IPC* 1303.9.5	b
15.	*IPC* 1303.9.5	d
CHAPTER 15		
1.	*UPC* 1210.3.3(1)	c
2.	*UPC* 1210.5	a
3.	*UPC* 1208.4.2	d
4.	*UPC* 1208.8.5.8.4(1)	c
5.	*UPC* 1208.7.1(2)	b
6.	*UPC* 1210.1.1	a
7.	*UPC* 1208.7.1(2)	b
8.	*UPC* 1210.9(5)	c
9.	*UPC* 1210.9(4)	a
10.	*UPC* 1211.3.1	d
11.	*UPC* 1211.3.2	c
12.	*UPC* 1213.3	b
13.	*UPC* 1211.5	d
14.	*UPC* 1213.3	c
15.	*UPC* 1213.3	b
16.	*UPC* 1213.3	d

—NOTES—

—NOTES—

—NOTES—